# PERSPECTIVES IN NONLINEAR DYNAMICS

# PERSPECTIVES IN
# NONLINEAR DYNAMICS

28 — 30 May 1985
Naval Surface Weapons Center

Editors:

**M F Shlesinger**
Office of Naval Research
**R Cawley**
Naval Surface Weapons Center
**A W Saenz**
Naval Research Laboratory
**W Zachary**
Naval Research Laboratory

World Scientific

*Published by*

**World Scientific Publishing Co Pte Ltd.**
**P. O. Box 128, Farrer Road, Singapore 9128**

Library of Congress Cataloging-in-Publication Data is available.

**PERSPECTIVES IN NONLINEAR DYNAMICS**

ISBN 9971-50-111-2
9971-50-114-7 pbk

QC133·P471 1986
#1792

Printed in Singapore by Kyodo Shing Loong Printing Industries Pte Ltd.

# INTRODUCTION

Dynamics, the oldest discipline of classical physics had its initial grand success with Newton's derivation of Kepler's laws. The clarity and beauty of classical dynamics went as far as to inspire a mechanistic clockwork view of the universe, which incidentally was not shared by Newton. This world view was over-turned with the advent of quantum mechanics with a God who plays dice with the universe. A joke making the rounds is "nonlinear dynamics is what physicists would have done 60 years ago if the computer had been invented and quantum mechanics had not".

Until recently classical mechanics was relegated to a first course in the graduate physics curriculum stressing its Hamiltonian formulation as a precursor to quantum mechanics. Mechanics was certainly not a popular field of active research. However, all of this has now changed and we are in the midst of a remarkable revolution which has vaulted dynamics into one of the most exciting and promising fields of science with consequences that cut across many disciplines. Nonlinear dynamics has been a unifying factor in science, helping to reverse the trend of specialization.

How, almost 300 years after the publication of Newton's *Principia* (1687) did dynamics once again reach the forefront of science? The answer lies in the shift from linear to nonlinear equations of motion. When compared to linear systems, nonlinear systems exhibit ergodic behavior or competition between nonlinear modes, they exhibit mode selection instead of mode superposition, effective reduction of degrees of freedom instead of completeness, relevant inherent geometries can be fractal instead of Euclidean, one can obtain solitons instead of waves, chaos instead of diffusion, and intermittency achieves a natural existence instead of being arbitrarily introduced as arising from noisy external sources.

A second part of the answer to the question lies in a historic reformulation of the harder fundamental problem of Newton's calculus: integration. Poincaré, Lie and Birkhoff fathered the global approach from which so many of the most important modern results have come, in which the problem of solving differential equations by effecting quadratures was replaced by the problem of asking for typical behaviors of the solution trajectories. Thus, two centuries after the publication of the *Principia* dynamics was geometrized. It was to take nearly another century before the computer, with assistance from Kolmogorov, Arnol'd and Moser, and from Smale, would make the study of complicated behavior accessible and exciting. The development of nonlinear dynamics has begun to take off now that experiment has joined this novel interplay between technology and theory. Nonlinear dynamics is truly a new calculus!

The emphasis has been placed on finding typical or universal properties of nonlinear systems. The progress in nonlinear dynamics has been so rapid that any Rip Van Winkle wandering into a modern dynamics conference could be startled not to find many landmarks. He would encounter with puzzlement a new vivid terminology such as chaos, strange attractors, fractal dimensions, fractal basic boundaries, solitons, crises, metamorphoses, KAM tori, horseshoes, devil's staircases, etc.

The Department of the Navy has recognized the importance of the new developments in dynamics in several ways. Research scientists at the Naval Research Laboratory and the Naval Surface Weapons Center formed in 1983 the Navy dynamics institute under partial sponsorship of the Office of Naval Research. The institute is involved with numerous research efforts, conducts weekly seminars, and organizes conferences. Under the sponsorship of the above three organizations, a workshop entitled "Perspectives in Nonlinear Dynamics," was held 28-30 May 1985 at the Naval Surface Weapons Center. In part, the lecturers were asked to provide the ONR Physics Division (which funds basic research in dynamics) with their views on what are the outstanding issues in nonlinear dynamics, their relative importance, and what types of research can resolve these issues. Their responses form the content of this book.

Broad areas covered include fluids, electrical systems, plasma, optics, mechanical systems, proteins, and mathematical and statistical approaches to investigating nonlinear dynamical systems.

We take this opportunity to thank all the lecturers and session chairmen for openly revealing their research plans in a forum attended by their keenest competitors. Their effort has been most helpful in helping the Navy chart new research directions in nonlinear dynamics.

PERSPECTIVES IN NONLINEAR DYNAMICS*

28-30 May, 1985

Naval Surface Weapons Center
White Oak, Silver Spring, Maryland

Morning Session Chairman:
Dr. Michael Shlesinger, Office of Naval Research, Arlington, Virginia

Welcoming address:
Dr. Lemmuel Hill, Technical Director, Naval Surface Weapons Center,
Dahlgren, Virginia

Introductory remarks:
Dr. David Nagel, Naval Research Laboratory, Washington, D.C.

Introductory remarks:
Dr. William Condell, Office of Naval Research, Arlington, Virginia

"Chaos in semi-conductors"
Prof. Robert Westervelt, Department of Physics, Harvard University,
Cambridge, Massachusetts.

"Experiments on nonlinear dynamics of solid state systems"
Prof. Carson Jeffries, Department of Physics, University of California,
Berkeley, California

_____

*Organized and sponsored by Office of Naval Research and the Navy Dynamics
Institute Program of the Naval Research Laboratory and the Naval Surface
Weapons Center.

"Fractals concepts in experiments on chaotic mechanical systems"
Prof. Francis Moon, Department of Theoretical and Applied Mechanics,
Cornell University, Ithaca, New York.

"Studies of nonlinear dynamics in lasers and other nonlinear optical systems"
Prof. Neal Abraham, Department of Physics, Bryn Mawr College, Bryn Mawr, Pennsylvania.

Moderated discussion (45 minutes)

Afternoon Session Chairman:
Dr. A. W. Saenz, Naval Research Laboratory, Washington, D.C.

"Atmospheric and oceanic models as dynamical systems"
Prof. Edward Lorenz, Department of Earth, Atmospheric and Planetary
Sciences, Massachusetts Institute of Technology, Cambridge, Massachusetts.

"Attractors for infinite dimensional systems"
Prof Jack Hale, Lefshetz Center for Dynamical Systems, Division of
Applied Mathematics, Brown University, Providence, Rhode Island.

"Nonlinear dynamical problems in channeling in crystals"
Prof. James Ellison, Department of Mathematics, University of New Mexico,
Albuquerque, New Mexico.

"Stability and prechaotic motion in Hamiltonian and nearly Hamiltonian systems"
Prof. Jerrold Marsden, Department of Mathematics, University of
California, Berkeley, California.

Moderated discussion (1 hour)

Wednesday, 29 May

Morning Session Chairman
Dr. William Caswell, Naval Surface Weapons Center, Silver Spring, MD

"Chaos and confusion in two-dimensional hydrodynamics"
Prof. Leo Kadanoff, James Franck Institute, University of Chicago,
Chicago Illinois.

"Interdisciplinary chaos"
Prof. Henry Abarbanel, Scripps Institution of Oceanography, University
of California, San Diego, California.

"Coherent structures and chaos in parital differential equations"
Prof. Alan Newell, Program in Applied Mathematics, University of
Arizona, Tucson, Arizona.

"Bifurcation and the integration of nonlinear differential systems"
Prof. Melvyn Berger, Department of Applied Mathematics, University of
Massachusetts, Amherst, Massachusetts.

Moderated discussion (1 hour)

Afternoon Session Chairman:
Dr. Woodford Zachary, Naval Research Laboratory, Washington, D.C.

"Computation theory, randomness and cellular automata"
Prof. Stephen Wolfram, School of Natural Sciences, Institute for
Advanced Study, Princeton, New Jersey.

"Computing/quantum chaos"
Dr. Bernardo Huberman, Xerox Corporation, Palo Alto, California.

"(Chaotic) ionization of atoms by microwave fields"
Prof. Peter Koch, Department of Physics, State University of New York,
Stony Brook, New York.

"Aspects of strange attractors in physical systems"
Prof. John Guckenheimer, Department of Mathematics, University of
California, Santa Cruz, California.

Moderated discussion (1 hour)

Banquet Lecture

"Zoe: superhealth for body and brain"
Arnold Mandell, M.D. Professor of Psychiatry, Laboratory of Biological
Dynamics and Theoretical Medicine, University of California, San Diego,
California

Abstract: The application will be demonstrated of some of the concepts
in mathematics and nonlinear dynamics to protein function, cardiovascular
physiology, polypeptide-endocrine regulation and brain behavior. An
attempt will be made to demonstrate the possibility that an ideal health
exists, can be characterized, facilitated, and its incipient loss
predicted.

Thursday, 30 May

Morning Session Chairman:
Dr. Robert Cawley, Naval Surface Weapons Center, Silver Spring, MD

"Quasi-periodicity and chaos"
Prof. Albert Libchaber, The James Franck and Enrico Fermi Institutes,
University of Chicago, Chicago, Illinois.

"Nonlinear Dynamics and chaos in oscillatory Rayleigh-Benard convection"
Dr. Robert Ecke, Los Alamos National Laboratory, Los Alamos, New Mexico.

"Low dimensional dynamics in high dimensional systems"
Prof. Harry Swinney, Department of Physics, University of Texas,
Austin, Texas.

"Nonlinear pattern formation from instabilities"
Prof. Jerry Gollub, Department of Physics, Haverford College,
Haverford, Pennsylvania.

Moderated discussion (1 hour)

Afternoon (Closing) Session Chairman:
Prof. James Yorke, Institute for Physical Science and Technology and
Department of Mathematics, University of Maryland, College Park,
Maryland.

"Effects of diffusion on islands in phase space in many dimensions"
Prof. Alan Lichtenberg, Electronics Research Laboratory, University of
California, Berkeley, California.

"Nonlinear dynamics beyond chaos"
Dr. Celso Grebogi, Center for Plasma Theory and Fusion Energy Research,
Department of Physics and Astronomy, University of Maryland, College
Park, Maryland.

"Nonlinear nights"
Prof. Robert H. G. Helleman, Theoretical Physics Center, Twente
University of Technology, Enschede, The Netherlands, and La Jolla
Institute for Nonlinear Problems, La Jolla, California.

Round Table Discussion, with additional moderators:

Prof Paul Linsay, Department of Physics, Massachusetts Institute of
Technology, Cambridge, Massachusetts.

Dr. Gottfried Mayer-Kress, Los Alamos National Laboratory, Los Alamos,
New Mexico.

Prof. Sheldon Newhouse, Department of Mathematics, University of North
Carolina, Chapel Hill, North Carolina.

CONTENTS

# PERSPECTIVES IN NONLINEAR DYNAMICS

# ATMOSPHERIC MODELS AS DYNAMICAL SYSTEMS

Edward N. Lorenz

Department of Earth, Atmospheric, and Planetary Sciences

Massachusetts Institute of Technology

Cambridge, MA   02139, U.S.A.

## ABSTRACT

We describe various types of approximation which have been introduced into the atmospheric equations to convert them into models. These models may be treated as dynamical systems. We examine one model in detail, and we enumerate some atmospheric problems where a nonlinear-dynamical approach might yield beneficial results.

## 1.   Introduction—models

The laws which govern the atmosphere may be expressed as a system of nonlinear equations. Deducing the typical behavior of the atmosphere from these equations constitutes a challenging problem in nonlinear dynamics. In attacking this problem one might expect to be guided by some of the recent studies in dynamical-systems theory, and one's first reaction to our title might be, "Why models of the atmosphere? Why not the real thing?" To understand our preference for models one needs to know what constitutes a dynamical system. One must also take a close look at the real atmosphere, and at the nature of the systems which comprise most atmospheric models.

We sometimes define a dynamical system as a finite system of coupled deterministically formulated ordinary differential equations in as many dependent variables [1]. Sometimes we relax the requirements to allow a countably infinite number of equations. Sometimes our systems consist of difference equations rather than differential equations. Whatever modifications we may permit, our interest is mainly in the long-term properties of typical solutions of the equations, rather than in methods of finding the solutions. We expect to encounter some special solutions, perhaps steady or periodic, whose properties differ considerably from those of most other solutions, but we expect that in some meaningful sense the special solutions will form a set of measure zero, so that their properties will not contribute to the overall average behavior.

What about the system of equations representing the laws which govern the atmosphere? Among these laws are the fundamental laws of hydrodynamics and thermodynamics, and we ordinarily take the attitude that they are known. A few details still elude us; for example, we do not know what determines just when a cloud, consisting of suspended water droplets or ice crystals, will release its water in the form of larger rain drops or snowflakes. Nevertheless, we are reasonably confident that a system obeying the atmospheric equations, as we have formulated them, will closely resemble the real atmosphere in its gross features and in many of its details.

What are typical solutions of these equations like? The equations are highly nonlinear, the most prominent nonlinear terms representing the quadratically nonlinear process of advection—the transport of momentum, heat, or moisture by the atmospheric motion. Any time-dependent solutions which we may be skillful or fortunate enough to discover by analytic procedures are likely to represent highly specialized behavior. In principle we can obtain typical solutions to any desired degree of approximation by numerical integration, although the actual task may be impractical. However, if our assumption regarding the exactness of the equations is correct, we can determine the nature of the typical solutions by observing the behavior of the atmosphere itself.

An outstanding characteristic of the atmosphere is the simultaneous presence of features of many spatial and temporal scales, and, in particular, many horizontal scales. There are globe-encircling westerly-wind currents, culminating in the jet streams. There are migratory vortices of subcontinental size, whose progression is responsible for many of the day-to-day weather changes in middle latitudes. There are tropical hurricanes, otherwise known as typhoons or tropical cyclones, which are less extensive but equally vigorous. There are intricately structured thunderstorms, comparable in size to

large mountains. There are fair-weather cumulus clouds, often no
larger than small hills. There are individual wind gusts, sometimes
only broad enough to sway a single tree at one moment. Our list is but
a sampling.

The above are not simply features which _may_ be present in a
correct solution of the equations; they are features which _must_ be
present in almost all time-dependent solutions. Any solution which
describes only the meanderings of a westerly current, or only the
progression of a chain of cyclonic and anticyclonic vortices, is a
special solution, belonging to the set of measure zero whose existence
we have noted.

It is evident that we lack the means for representing, even at a
single instant, global fields of wind, temperature, and moisture which
contain several thousand thunderstorms and hundreds of millions of
gusts. In short, we are limited by the speed and capacity of today's
most powerful computers, or of our brains, from determining typical
solutions of the most realistic atmospheric equations which we can
formulate. As a dynamical system the real atmosphere does not lend
itself to convenient investigation.

In view of these limitations, how is it possible for dynamic
meteorology, which was actually a well-established discipline long
before the advent of computers, to accomplish anything? Several lines
of pursuit are available.

We may use the equations, without actually solving them, to study
various atmospheric phenomena and processes. For example, we may
derive from the exact equations an expression for the time derivative
of the total energy of the vortices, and we may identify the various
terms in the expression with particular physical processes. If
adequate observational data are available, we may then evaluate the
long-term averages of the various terms, and learn which physical
processes play leading roles.

Alternatively, we may introduce various approximations. A common
procedure consists of linearizing the equations. The great advantage
of linear systems, aside from relative ease of solution, is
superposability of solutions. Thus, we may find solutions in which all

features have 3000-kilometer wave lengths, and others in which they all have 3-kilometer wave lengths, and we should then be able to study large-scale vortices and cumulus-cloud circulations independently of one another. Of course, any direct influence of one feature on the other will be suppressed.

With the advent of computers, numerical methods of solution have become popular, although many dynamicists still find the earlier procedures more appealing. As we have seen, essentially exact numerical solutions of the exact equations are unattainable, and again we must introduce various approximations, but there is no need to remove the nonlinearity.

Along with the adoption of numerical techniques has come a change of perspective. Whereas we were previously content to find approximate solutions of the equations governing the atmosphere, we now take the attitude that we are finding exact solutions of models of the atmosphere. In short, our atmospheric models are simply systems of equations, derived by introducing various approximations into the exact equations, and arranged so as to be amenable to analytic or numerical integration. When the model is to be handled numerically, we may regard the finite-differencing scheme, and even the round-off procedure, as a part of the model. This contrasts with the situation in some fields, where models are often constructed simply by postulating interrelations among various features.

Many types of approximation are in common use [2]. First, we may simplify the physical nature of the atmosphere or its surroundings. For example, we often ignore the presence of gaseous, liquid, and solid water, and treat the atmosphere as an ideal gas. This step appears to handle the largest-scale features reasonably well, although it would be fatal in dealing with phenomena like tropical hurricanes, which depend upon water for their origin and maintenance. Likewise, we often replace the spherical surface of the earth by an infinite or bounded plane. We represent the effect of the earth's rotation by a force--the Coriolis force--which deflects the wind to the right in the northern hemisphere and to the left in the southern. We assume that features which would develop under such conditions are qualitatively like those which do develop above a rotating spherical earth. The globe-encircling westerly current, for example, would become rectilinear, but its vertical and cross-latitude structure might remain virtually unchanged. In conjunction with the latter simplification, we often omit the earth's topographic features.

We may instead modify or eliminate certain physical processes. If we replace the equation of vertical motion by the hydrostatic approximation, which balances gravity against the vertical pressure force, and equates the pressure at a point to the weight of a column of air extending upward from that point, we obtain a considerably simpler system which is incapable of propagating sound waves, but is scarcely distinguishable from the exact equations in its treatment of systems larger than thunderstorms. If we also replace one equation of horizontal motion by the geostrophic approximation, which balances the Coriolis force against the horizontal pressure force, and effectively equates the pressure to a stream function for the wind, so that a low pressure center and a cyclonic vortex become equivalent, we obtain a still simpler system which is incapable of propagating inertial-gravity waves, but handles the largest-scale atmospheric features fairly well outside of the tropics. A combination of the hydrostatic and geostrophic approximations equates the vertical shear of the wind to the horizontal temperature gradient. Sometimes we simply discard annoying terms from an equation with little regard for their physical meaning.

Further approximations are necessary if numerical methods of solution are to be used. The exact equations, and the equations obtained from them by introducing various physical simplifications, are generally formulated as a set of partial differential equations. If radiative heating and cooling enter explicitly, the equations will also contain integrals. Before the usual numerical procedures can be carried out, the field of each dependent variable must be represented by its values at a finite grid of points, and finite differences and sums must replace derivatives and integrals. Alternatively each variable may be expressed as a series of spherical harmonics or other orthogonal functions; multiple Fourier series may be used if plane geometry has been introduced. The equations are then transformed into a countably infinite system of ordinary differential equations, with a countably infinite number of terms in each equation. Again, all but a finite number of equations, and all but a finite number of terms in each equation, must be discarded before numerical integration can begin.

If we intend to use our model to study the smaller scales, we can resolve these scales by restricting the model to a limited area. We may include the influence of larger-scale features through prescribed boundary conditions. If instead our purpose is to study the larger scales, we must discard the small scales. However, the small scales influence the large scales; the circulation within each cumulus cloud, for example, can carry significant amounts of heat and moisture to higher levels. We are therefore well advised to include additional terms in our model, representing the probable influence of an extensive ensemble of small-scale features.

How well do models with both physical and mathematical simplifications perform? Many of them have been constructed for the purpose of weather forecasting. These typically contain thousands of equations; the chief limitation to their size is the speed and capacity of the computers which are compatible with the budgets of the various weather services. The forecasts produced by the largest models, with several hundred thousand variables, compare favorably with forecasts produced by other means, although they are far from perfect.

Models used primarily for research are sometimes equally large, but, when only qualitatively correct results are desired, they are often made much smaller. The most highly simplified models are the "low-order models", which are often designed to study specific phenomena, and where, ideally, one retains the minimum amount of physics and the minimum resolution needed to describe the phenomenon of interest [3]. Low-order models typically have fewer than a hundred variables, and sometimes fewer than ten. Not surprisingly, some of the lowest-order atmospheric models have become some of the most intensively studied dynamical systems.

In the following paragraphs we shall describe how a dynamical-systems approach may be applied to a specific model. We shall then enumerate several problems where this approach may yield beneficial results.

## 2. A simple model

For a model which is readily treated as a dynamical system, we choose what is perhaps the simplest set of equations which can make some claim to being a model of the global atmospheric circulation [4]. This low-order model contains only three ordinary differential equations:

$$dX/dt = -(Y^2 + Z^2) - a(X - F) , \qquad (1)$$

$$dY/dt = XY - bXZ - (Y - G) , \qquad (2)$$

$$dZ/dt = bXY + XZ - Z . \qquad (3)$$

In Eqs. (1)-(3), X represents the intensity of the middle-latitude westerly wind current in the northern or southern hemisphere; the two hemispheres may be treated as mirror images of each other. Simultaneously, X represents the cross-latitude temperature gradient in either hemisphere. The wind and temperature fields are assumed to be in permanent geostrophic balance, so that a single variable can represent both. We shall refer to these combined fields as the zonal flow, using the term "flow", as we often do in meteorology, to refer not only to the motion field but also to the pressure and temperature fields which must accompany it. The horizontal and vertical structures of the zonal flow are prespecified, and only the intensity is allow to vary.

The variables Y and Z represent the cosine and sine phases of a chain of vortices superposed on the zonal flow. The horizontal and vertical structures of the vortices are prespecified, and only their longitude and intensity are allowed to vary. Relative to the zonal flow, the vortices are scaled so that $X^2 + Y^2 + Z^2$ is proportional to the total (kinetic plus potential plus internal) energy.

The vortices are linearly damped by viscous and thermal processes, and the damping time for the vortices is chosen as the time unit. The constant a is the reciprocal of the damping time for the zonal flow, and we let a < 1. In interpreting our results we let one time unit equal five days.

The vortices are constrained to tilt westward with increasing elevation, whence, under the assumed geostrophic balance, the poleward-moving air is warmer than the equatorward-moving air at the same latitude, and the net effect of the vortices is to transport heat poleward, thus reducing the temperature gradient. This effect accounts for the terms $-(Y^2 + Z^2)$ in Eq. (1). At the same time the transport does not alter the total energy of the atmosphere, so that the energy extracted from the zonal flow must be absorbed by the vortices; hence the terms XY in (2) and XZ in (3). The variables are scaled to make the coefficients of these terms equal to unity.

In addition to strengthening the vortices, the zonal flow transports them eastward (or westward, if X < 0). The constant b measures the ratio of the transport rate to the amplification rate, and we assume that b > 1.

The principal external driving force--the contrast between the low-latitude and high-latitude solar heating--acts directly on the zonal flow, and is represented by aF. A secondary driving force, which varies with longitude, and may be assumed to depend upon the contrasting thermal properties of the oceans and continents, acts on the vortices, and is represented by G.

In view of the simplicity of the included physical processes, it is evident that we could have constructed essentially the same model by simply postulating relationships among the variables, as is commonly done in some other disciplines. Actually, however, the model may be derived through systematic simplifications of the exact equations, including omission of moisture, introduction of the hydrostatic and geostrophic approximations, and extreme truncation.

As a dynamical system, Eqs. (1)-(3) possess a rich bifurcation structure, and the solutions exhibit many forms of behavior as the four parameters are altered. In this description we shall confine our attention to the fixed values a = 1/4, b = 4, and F = 8, and examine the changes in behavior as G increases from zero. Changing the sign of G has no effect other than changing the signs of Y and Z.

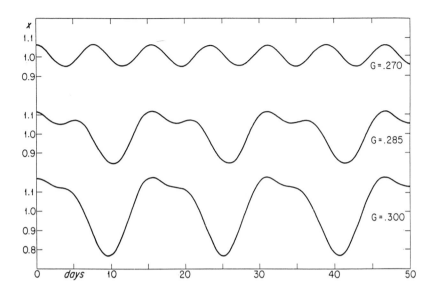

Fig. 1. Variations of X in the stable periodic solution of Eqs. (1)–(3), for G = 0.270, 0.285, and 0.300. In each case a = 0.25, b = 4.0, and F = 8.0.

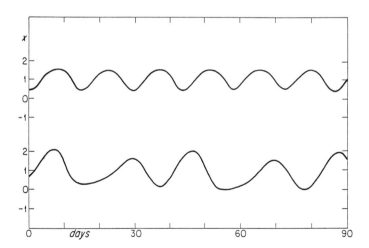

Fig. 2. Variations of X in the two distinct stable periodic solutions of Eqs. (1)–(3), for G = 0.8. As in Fig. 1, a = 0.25, b = 4.0, and F = 8.0.

When G = 0, the equations possess the single steady solution
X = F, Y = 0, Z = 0, representing undisturbed eastward flow. This is
seen to be unstable if F > 1, in which case there is a periodic
solution X = 1, Y = R cos bt, Z = R sin bt, with $R^2$ = a(F − 1), representing steadily progressing vortices. By transforming Y and Z to R
and θ, where tan θ = Z/Y, we readily see that this solution is stable.

When G acquires a small positive value, we may expect a modified
periodic solution in which the vortices tend to intensify when Y > 0
and weaken when Y < 0. A resultant effect of these variations of R
will be oscillations of a similar period in the zonal flow X; these in
turn will produce additional variations in the behavior of the
vortices.

Explicit solutions when G > 0 may be sought numerically. We find
that the anticipated behavior does occur until G reaches 0.277, when
the solution becomes unstable. The bifurcation at this value at first
resembles a classical period-doubling bifurcation [5], with
oscillations occurring at the original frequency, but with weaker
oscillations alternating with stronger ones. However, when G reaches
0.294 the weaker maximum disappears, at least in the variations of X,
and the frequency has indeed been halved. Fig. 1 compares the
variations of X for G = 0.27, 0.285, and 0.30.

No further doublings in this solution are evident, but at G = 0.75
a new periodic solution is born, and there are two disjoint attractors.
Fig. 2 shows the variations of X for the two periodic solutions, for
G = 0.8.

The new solution soon enters a period-doubling sequence, and
becomes chaotic when G reaches 0.85, but the older solution remains
stable, although its basin of attraction becomes increasingly smaller,
up to G = 0.99, when it is swallowed up by the chaotic solution. For
most values of G from 0.99 to 1.367, where a new stable steady solution
appears, there is a single strange attractor, but, within this range of
G, there are some intervals, notably near G = 1.19, where the solution
is periodic. Such periodic windows are common occurrences in systems
containing very few variables, and are probably rarer in more detailed
atmospheric models.

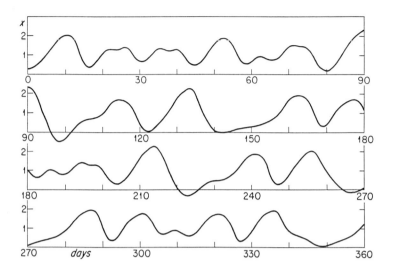

Fig. 3. Variations of X in an aperiodic solution of Eqs. (1)-(3), for G = 1.1, during a one-year interval, shown as four consecutive 90-day segments. As in Fig. 1, a = 0.25, b = 4.0, and F = 8.0.

Fig. 3 shows the variations of X during a one-year interval, when G = 1.1, displayed as four consecutive 90-day segments. The lack of periodicity is apparent, and there is some tendency to switch back and forth between weaker more rapid oscillations, like those in the upper curve in Fig. 2, and stronger less rapid oscillations, characteristic of the lower curve.

Fig. 4 shows the intersection of the attractor with the plane Z = 0, when G = 1.1. An intricate structure is evident. Qualitatively the appearance of the attractor is about the same on either side of a periodic window, and it contrasts with the small collection of points which would replace Fig. 4 in the window. Fig. 5 shows one half of the intersection of the attractor with the plane Y = 0, with higher resolution, and many details, including the interior blank areas, are more easily seen.

What does this analysis tell us about the real global atmospheric circulation? It certainly does not reveal what processes maintain the zonal flow and the vortices; at most it indicates that the processes which we have included in the model may be capable of doing so. It

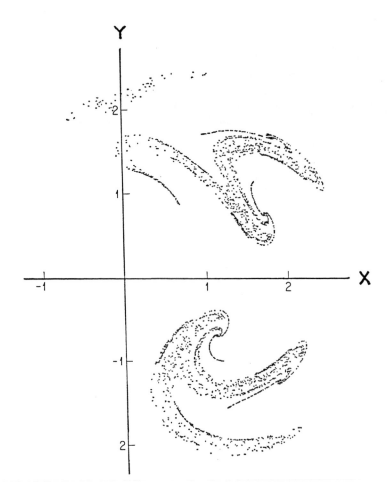

Fig. 4.  The intersection of the attractor of Eqs. (1)-(3), for
the conditions of Fig. 3, with the plane Z = 0, as represented by
3000 successive crossings of a single orbit.

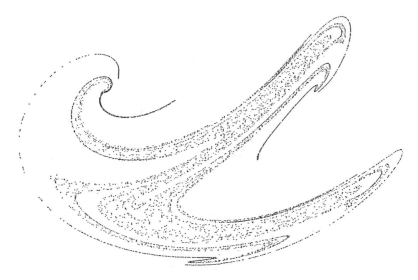

Fig. 5. A portion of the intersection of the attractor of Eqs.
(1)-(3), for the conditions of Fig. 3, with the plane Y = 0, as
represented by 12000 crossings of a single orbit.

does not, for example, imply that a transfer of energy from the zonal
flow, rather than a transfer from smaller-scale features, is the
process which maintains the vortices, since the model is incapable of
saying anything about the smaller-scale features.

Perhaps the property of the model's behavior which most closely
resembles its real atmospheric counterpart is the erratic variations of
the zonal flow, as displayed in Fig. 3. Various conflicting
explanations for such variations in the real atmosphere have been
offered. The model tells us that there is no certainty that variable
external activity is involved; in the model everything external is
steady. Moreover, complicated internal mechanisms need not be
involved; everything in the model is simple.

In the real atmosphere vortices seem to be always present; the
flow never becomes purely zonal. One sometimes assumes, however, that
the flow would become zonal, at least temporarily, if one waited long
enough, and, in attempting to explain certain occurrences, one
sometimes feels compelled to explain how they could have evolved from
an essentially zonal state, or perhaps even from a state of rest. In

our model the attractor clearly excludes a cylinder surrounding the X-axis, where the energy of the vortices is small. This suggests that, in the real atmosphere, states which are nearly zonal may never be approached, much less attained. The frequently used initial conditions in numerical integrations of various model equations, consisting of a zonal flow plus a small perturbation, may therefore be illogical, even if convenient.

3. Conclusion--atmospheric problems

In looking at Eqs. (1)-(3) as a dynamical system, we have gained some insight into one atmospheric problem--the coupling of the zonal flow and the large vortices--but the simplicity of the model has prevented us from treating all aspects of the problem. We could have learned more, for example, by using a model which does not assume that the vortices transport heat poleward, and instead determines its own heat transport, or a model which does not presuppose geostrophic balance, and instead produces its own balance. Meanwhile, there are numerous other atmospheric problems which presumably can be profitably investigated by constructing appropriate models and treating them as dynamical systems.

Probably the classical example of irregular or chaotic behavior is turbulent behavior. Atmospheric turbulence is especially complicated because of its inhomogeneity and intermittency. In the lowest few meters of the atmosphere, where most of us spend most of our lives, the vertical extent of a turbulent eddy is limited by the proximity of the ground or the sea, but no such limitation exists farther aloft. Turbulence covers a wide range of scales, and even the largest vortices possess some of the properties of anisotropic turbulent eddies. The volume of research in atmospheric-turbulence theory has been enormous, but many basic questions remain unanswered, and it has recently been predicted that future research will involve the concepts of strange attractors and coherent structures [6].

Coherent structures, which pose another problem, are in a sense the antithesis of turbulence, but they are equally nonlinear. They consist of features which retain their form over extended time intervals, even though they may be superposed on an essentially turbulent background. The classical example is the soliton. Coherent atmospheric structures appear to be more easily found in models than in the atmosphere itself [7], but there is some evidence that tropical hurricanes are structures of this sort.

A more specialized problem is initialization. This is one step in the process of numerical integration of operational weather-forecasting models which do not preassume geostrophic balance, and it is needed because small errors in wind and temperature observations usually produce observed initial states where the geostrophic unbalance is much greater than in the true initial state. Large-amplitude inertial-gravity waves then ensue, and contaminate the forecast. Initialization procedures attempt to replace the observed initial state by a slightly different state which is in approximate geostrophic balance, and will remain in approximate geostrophic balance as the forecast evolves. A number of initialization procedures have been developed [8], but even the more successful ones are often awkward to apply. As a problem in dynamical-systems theory, initialization may be equated with seeking a special invariant manifold, sometimes called the "slow manifold" [9]. The problem has been approached via invariant-manifold theory [10].

We close with a somewhat more detailed discussion of another problem--predictability. This concerns the extent to which it is possible to predict various aspects of the weather at various ranges. The limiting factor is the sensitive dependence of atmospheric models, and presumably of the real atmosphere, on initial conditions; our observations will not distinguish among a number of nearly identical states, and, since these states will develop differently, there will be no basis for choosing among a number of considerably different future states.

The key quantity is the rate at which small differences between states will amplify, traditionally expressed in terms of a doubling time. For a simple system like Eqs. (1)-(3), this is proportional to the reciprocal of the largest Lyapunov exponent. One might suppose that a similar relationship would hold in more general models, but actually this is not the case, if the model has sufficiently high resolution.

The reason is again the abundance of scales found in the atmosphere. Errors in observing smaller-scale features, especially the more energetic ones, will grow rapidly; the error in observing the details of a thunderstorm, if such observations are indeed performed at all, should amplify at least as rapidly as the thunderstorm itself, doubling in half an hour or less. In short, the largest Lyapunov exponenets of high-resolution atmospheric models, and of the atmosphere itself, are associated with small scales.

These same errors, however, soon acquire limiting amplitudes, at a time when the errors in the larger scales are just beginning to reveal their growth. The latter errors, aside from growing more slowly and therefore being associated with smaller Lyapunov exponents, continue to grow much longer, generally doubling in two days or more [11], and they ultimately acquire much larger amplitudes than the small-scale errors; they therefore provide the major contribution to the total error in the forecast.

We need to know, then, not only how rapidly small-amplitude errors of all scales will depart from their initial magnitudes, but also how slowly large-amplitude errors will approach their limiting magnitudes. Such information is especially pertinent to extended-range prediction and the prediction of climate. We also need to know how errors in separate weakly coupled scales will influence each other. Some relevant individual studies have been performed [12], but the development of a coherent theory presents an especially challenging problem in nonlinear dynamics.

Acknowledgment. This work has been sponsored by the Climate Dynamics Program of the Atmospheric Science Section, National Science Foundation, under Grant 82-14582 ATM.

## REFERENCES

1. Guckenheimer, J., and Holmes, P., 1983: Nonlinear oscillations, dynamical systems, and bifurcations of vector fields. Springer-Verlag, New York, 453 pp.

2.  Holton, J. R., 1979: An introduction to dynamic meteorology, 2nd ed. Academic Press, New York, 391 pp.

3.  Lorenz, E. N., 1982: Low-order models of atmospheric circulations. J. Meteorol. Soc. Japan, 60, 255-267.

4.  Lorenz, E. N., 1984: Irregularity: a fundamental property of the atmosphere. Tellus, 36A, 98-110.

5.  Feigenbaum, M. J., 1978: Quantitative universality for a class of nonlinear transformations. J. Stat. Phys., 19, 25-52.

6.  Tennekes, H., 1985: A comparative pathology of atmospheric turbulence in two and three dimensions. Turbulence and predictability in geophysical fluid dynamics and climate dynamics, M. Ghil, R. Benzi, and G. Parisi, eds., Italian Phys. Soc., North Holland Pub. Co., Amsterdam, 45-70.

7.  Malanotte Rizzoli, P., 1984: Coherent structures in planetary flows as systems endowed with enhanced predictability. Predictability of fluid motions, G. Holloway and B. J. West, eds., Amer. Inst. Physics, New York, 223-245.

8.  Daley, R., 1981: Normal mode initialization. Rev. Geophys. and Space Sci., 19, 450-468.

9.  Leith, C. E., 1980: Nonlinear normal mode initialization and quasi-geostrophic theory. J. Atmos. Sci., 37, 958-968.

10. Kopell, N., 1985: Invariant manifolds and quasi-geostrophic theory. Physica, 14D, 203-215.

11. Lorenz, E. N., 1982: Atmospheric predictability experiments with a large numerical model. Tellus, 34, 505-513.

12. Lorenz, E. N., 1984: Estimates of atmospheric predictability at medium range. Predictability of fluid motions, G. Holloway and B. J. West, eds., Amer. Inst. Physics, New York, 133-139.

# PERSPECTIVES ON STUDIES OF CHAOS IN EXPERIMENTS

Harry L. Swinney

Physics Department

The University of Texas

Austin, Texas 78712

## ABSTRACT

Noisy signals have traditionally been analyzed using Fourier spectra, correlation functions, and other methods developed for linear systems. However, these techniques are inadequate for characterizing chaos in deterministic nonlinear systems. More appropriate properties for characterizing chaos are the Lyapunov exponents, entropy, and dimension of the attractors. Methods are now being developed to compute these properties from time series data obtained in laboratory experiments. We briefly describe some of these methods and their limitations and applications.

1.    Attractors from Time Series

In the analysis of most experiments the first step is the reconstruction of multidimensional phase space portraits from measurements of a single (or, at most, a few) observables. A vector time series $X(t)$ is obtained from a scalar time series $s(t)$ using time delays:[1] $X(t) = \{x_0(t), x_1(t), ..., x_n(t), ...\}$, where $x_n = s(t+nT)$. The time delay T is in principle arbitrary for an *infinite* amount of *noise-free* data. However, experimenters have found that for real data the amount of information that can be extracted from the reconstructed attractors depends strongly on the choice of T. Recently Fraser[2-4] has shown that the mutual information function from information theory provides a quantitative criterion for choosing T. An algorithm for computing the mutual information was developed, tested on model systems, and applied to experimental data.[2-4] The method works well for 2 or 3 dimensional systems, but has not yet been tested for higher dimensional systems.

2.    Characterizing Chaos

In this section we mention briefly some methods that have been proposed for computing some of the dynamical invariants (Lyapunov exponents, entropy, dimension) that characterize chaotic attractors.

Two methods have been proposed for estimating the nonnegative Lyapunov exponents of attractors constructed from a time series. In a method developed by Wolf et al.[5] the largest exponent is determined by monitoring the separation between a point on a fiducial attractor and a nearby point. When the separation becomes large, a new point is found near the fiducial trajectory with the constraint that the new point must lie as nearly as possible in the same direction as the point it replaces. The procedure continues until the fiducial trajectory is followed to the end of the time series. Then the largest Lyapunov exponent is given by

$$\lambda_1 = (1/T) \sum_i \log (L'_i/L_i), \tag{1}$$

where T is the experiment duration, $L'_i$ is the length evolved from a length $L_i$, and the summation is over the number of replacement steps. The sum of the two largest exponents, $\lambda_1 + \lambda_2$, can be determined by following the evolution of areas $A_i$ instead of lengths $L_i$. In principle the method could be extended to additional exponents by following the evolution of 3-volumes, etc., but in practice the approach becomes unwieldy for more than two exponents.

Eckmann and Ruelle[6] have suggested another method for computing Lyapunov exponents that in principle should yield all the positive exponents. Least-squares estimates of the Jacobian at each point are obtained by following the evolution of groups of nearby points for a short time. The Lyapunov exponents are then obtained from the eigenvalues of the products of the estimated Jacobians.

The Wolf method for determining $\lambda_1$ has been tested on model systems and on laboratory data, and the exponent estimates have been found to be robust for reasonable variations of fitting parameters (the maximum of the evolved lengths, $\epsilon$; the evolution time, $\Delta t$; the embedding dimension, m; and the number of data points, N). The Jacobian estimation method is similarly fairly insensitive to variations in $\epsilon$, $\Delta t$, and N; however, Vastano and Kostelich[7-8] have found that the results of this method depend *strongly* on the embedding dimension m. Therefore, since m is not known *a priori*, the Jacobian method does not appear to be useful for the analysis of experimental data, despite claims to the contrary.[9]

For systems with several positive Lyapunov exponents the metric (Kolmogorov-Sinai) entropy[10] is probably of greater interest than the largest Lyapunov exponent. The metric entropy, which is conjectured to be equal to the sum of the positive Lyapunov exponents,[11] is the long-time average rate at which information about the system state (known from previous measurements) is lost due to the system dynamics. Calculation of entropy from the definition requires the determination of the probability $P(S_n)$ of observing any possible symbol sequence $S_n=\{s_1,s_2,...,s_n\}$ arising from a generating partition of the phase space.[12-14] For most physical systems it is conjectured that any reasonably fine partition will be a generating partition.

Termonia[15] has presented an algorithm for computing the entropy directly from the definition, but unfortunately the amount of data required appears to be quite large, even for low dimensional systems. Recently Fraser[3-4] has developed what appears to be a more efficient technique for obtaining entropy estimates from the definition. The key feature of Fraser's method is that the phase space is partitioned into boxes whose sizes are tailored to local conditions. The method seems promising, but has not yet been fully implemented and tested.

Grassberger and Procaccia[16] have proposed an indirect method for estimating the order-2 Renyi entropy, which provides a lower bound on the metric entropy. The data requirements (which are the same as for computing the correlation dimension[21]) are less demanding than for the Termonia method. However, the Grassberger-Procaccia method has not yet been widely tested or used, and an attempt to apply the method to turbulent Couette-Taylor flow data was not successful.[22]

Several methods have been proposed for computing the fractal dimension or capacity D (a simplified version of the Hausdorff dimension), the information dimension d, and the correlation dimension $\nu$ of attractors.[17-24] These quantities are related by $\nu \leq d \leq D$, but often equality or near equality is found to hold. (See Farmer et al.[20] for a discussion of the various dimensions and the relations between them.) In principle the fractal dimension can be computed quite straightforwardly from the definition: the phase space is divided into cubes of size r and the number of cubes $N(r)$ that contain a part of the attractor is determined; D is given by $N(r)\sim r^{-D}$ for small r. In practice, however, the number of cubes required increases exponentially with D, and computation of D greater than about 3 is not possible.[18]

A more efficient approach to computing D has been proposed by Badii and Politi.[19] In their method the average distance between nearest neighbors is determined as a function of the total number of data points. This algorithm seems to be about as efficient as the Grassberger-Procaccia[21] algorithm for computing $\nu$ and the algorithm of Farmer et al.[20] for computing d. Dimension values as large as about six have been obtained by these methods. Recently Somerjai and Ali[23] have suggested that a nearest neighbor algorithm introduced by Pettis et al.[24] for the analysis of patterns could be used, if the results were appropriately "calibrated", to determine dimensions as large as perhaps 20. A careful examination of this intriguing possibility is warranted.

From this brief discussion it should be clear that the characterization of chaos is in its infancy. One problem is that almost any algorithm developed to determine dimension, entropy, Lyapunov exponents, etc. will yield a number, but it is often *very* difficult to establish whether or not that number correctly characterizes the dynamics of the physical system. There have been no systematic studies of the requirements on the quantity and accuracy of the data needed to determine the different dynamical invariants. How does the number of data points needed scale with dimension? Are the algorithms robust under variations of fitting parameters? In particular, how do the results depend on embedding dimension? Will it ever be possible to determine with any confidence the properties of high dimensional $(d > 10)$ attractors? A thorough numerical and analytical examination of such issues is crucial for progress in the characterization of chaos.

3.    Applications and Future Work

The development of techniques for computing different dynamical invariants opens up many possibilities for quantitative comparisons between theory and experiment. For example, beyond the accumulation point of the period doubling sequence, the largest Lyapunov exponent is predicted to be described by[25]

$$\lambda_1 \sim (\mu - \mu_c)^\beta, \tag{2}$$

where $\mu$ is the bifurcation parameter and $\beta$ is a universal number, 0.4498069... . A few other routes to chaos besides period doubling have been identified in theory and experiment. Are there many more? Are there universal scaling laws that describe the asymptotic behavior of the metric entropy, Lyapunov exponents, and dimension near critical points? What is the role of symmetry in determining the routes to chaos? Are the routes to chaos different in low dimensional and high dimensional systems? What is the role of system size (or aspect ratio, which is the ratio of the largest to smallest dimension of a system)? How can the concept of strange attractors be applied to open systems such as boundary layers or mixing layers, where the Reynolds number is a function of position?

Future work should focus particularly on those situations where direct comparison can be made between experiments and the equations of motion. Of course the behavior of nonequilibrium systems can in general be highly complex; systems of interest typically have several independent control parameters, and even for the same constraints there are multiple basins of attraction. Studies of higher order critical points ("co-dimension n" points), which are the intersections of two or more transition boundaries in the space of control parameters, should yield insight into the kinds of transitions between different dynamical regimes that can occur. The behavior in the neighborhood of codimension-n points can be analyzed in terms of normal forms (amplitude equations), taking into consideration the relevant symmetries, and experiments can be designed to examine behavior along different paths approaching these points. Experiments with periodic and nonperiodic (noisy) forcing will be important in studies of codimension-n bifurcations, and experiments designed to alter a system's symmetry will help elucidate the role of symmetry breaking.

Nonlinear dynamics research has been concerned thus far primarily with *temporal* instabilities and chaos. Clearly future studies must examine the *spatial* as well as temporal behavior of systems described by partial differential equations. Methods must be developed for characterizing spatial chaos and for making simultaneous measurements at many spatial points. Flow visualization with digital imaging is a step in the right direction, but methods should be developed to make closer contact with the dynamical variables of theory (e.g., the Eulerian velocity and vorticity fields). An understanding of the development and evolution of spatial patterns is important in biology and chemistry (reaction-diffusion systems) as well as in viscous fluids (Navier-Stokes systems). Therefore, the techniques that are now being developed to study spatial coherent structures in fluid dynamics should find immediate application in chemistry and biology, where well-controlled studies of patterns have been rare. This is a fertile area for future research.

Acknowledgements

This research is supported by the National Science Foundation Fluid Mechanics Program and the Department of Energy Office of Basic Energy Sciences. The work on quantifying chaos is conducted in collaboration with A. Brandstater, A. Fraser, E. Kostelich, J. Swift, J. Vastano, and A. Wolf.

References

1.  J.C. Roux, R.H. Simoyi, and H.L. Swinney, Physica **8D**, 257 (1983).
2.  A. Fraser and H.L. Swinney, Phys. Rev. A., to appear.

3. A. Fraser, in *Entropies and Dimensions*, ed. by G. Meyer-Kress (Springer, (1986).
4. A. Fraser, in preparation.
5. A. Wolf, J. Swift, H.L. Swinney, and J. Vastano, Physica **16D**, 285 (1985).
6. J.P. Eckmann and D. Ruelle, Rev. Mod. Phys. **57**, 617 (1985).
7. J. Vastano and E. Kostelich, submitted to Phys. Rev. Lett.
8. J. Vastano and E. Kostelich, in *Entropies and Dimensions*, ed. by G. Meyer-Kress (Springer, 1986).
9. M. Sano and Y. Sawada, Phys. Rev. Lett. **55**, 1082 (1985).
10. A.N. Kolmogorov, Dokl. Nauk SSSR **124**, 754 (1959); Ya. Sinai, Dokl. Akad. Nauk SSSR **124**, 768 (1959).
11. Ya. B. Pesin, Usp. Mat. Nauk. **32**, 55 (1977). [English translation: Math. Surveys **32**, 55 (1977)]
12. D. Ruelle, Comm. Math. Phys. **55**, 47 (1977).
13. J.P. Crutchfield and N.H. Packard, Physica **7D**, 201 (1983).
14. R. Shaw, *The Dripping Faucet as a Model Chaotic System* (Aerial Press, Santa Cruz, CA, 1985).
15. Y. Termonia, Phys. Rev. **A29**, 1612 (1984).
16. P. Grassberger and I. Procaccia, Phys. Rev. **A28**, 2591 (1983).
17. H. Froehling, J. Crutchfield, J.D. Farmer, N. Packard, and R. Shaw, Physica **3D**, 605 (1981).
18. H. Greenside, A. Wolf, J. Swift, and T. Pignataro, Phys. Rev. **A25**, 3453 (1982).
19. R. Badii and A. Politi, Phys. Rev. Lett. **52**, 1661 (1984).
20. J.D. Farmer, E. Ott, and J. Yorke, Physica **7D**, 153 (1983).
21. P. Grassberger and I. Procaccia, Physica **9D**, 189 (1983).
22. A. Brandstater, private communication.
23. R.L. Somerjai and M.K. Ali, "An efficient algorithm for estimating dimensionalities," preprint (1985).
24. K. Pettis, T. Bailey, A. Lain, R. Dubes, IEEE Trans. Pattern Analysis & Machine Intelligence PAMI-1 25 (1979).
25. B.A. Huberman and J. Rudnick, Phys. Rev. Lett. **45**, 154 (1980).
26. H.L. Swinney, Physica **7D**, 3 (1983).

# PATTERN FORMATION FROM INSTABILITIES

J.P. Gollub

Physics Department, Haverford College, Haverford, PA 19041
Physics Department, University of Pennsylvania
Philadelphia, PA 19104

## ABSTRACT

We briefly review several recent experiments illustrating
processes that control nonlinear pattern formation from
hydrodynamic instabilities. These processes include a
secondary instability, spatial forcing, and defect dynamics. We also show that the interaction between
distinct spatial patterns can produce chaos.

## 1. Introduction

The development of complex spatial structures through bifurcations that break spatial symmetries is a challenging problem in
nonlinear dynamics. Much of the current interest in chaos was stimulated by hopes of understanding the phenomenon of turbulence in
fluids and analogous problems in plasmas, chemical reactors, lasers,
etc. Chaotic behavior similar to that shown by simple nonlinear models
with a few degrees of freedom has been demonstrated in these systems.[1] However, it has become clear that turbulent phenomena
frequently require models with many degrees of freedom. Experiments
capable of testing such models are still in their infancy.

This paper is concerned with spatial structures, especially
those that are static or slowly varying. We review some recent experimental work performed at Haverford College and The University of
Pennsylvania on spatial patterns in hydrodynamic systems near the
thresholds of instabilities. These studies are a small first step toward
the more ambitions goal of understanding phenomena that are chaotic in
both space and time. We begin by posing some significant questions

about the formation of nonlinear spatial structures.

First, we would like to understand the problem of pattern selection. When a periodic spatial structure (for example, an array of convective rolls) arises as a result of an instability, there are generally many possible structures (a continuum of wavevectors, for example) that are consistent with stability considerations. However, the set of patterns that are actually realized experimentally as stationary states is much smaller. We describe several processes that are responsible for pattern selection: secondary instabilities, external periodic forcing, and boundary and defect dynamics. Then we consider the question of the relationship between spatial patterns and chaotic phenomena. We show that the competition between different spatial patterns can lead to chaos.

2.  Pattern Selection near the Onset of Convection: The Eckhaus
    Secondary Instability [2]

The Eckhaus instability is a general mechanism of pattern selection that occurs in any translationally invariant system where a normal bifurcation produces a spatially periodic structure in one space direction.[3] The linear stability curve for the primary instability and the stability boundary of the Eckhaus secondary instability are parabolas that are tangent at the minimum. If the initial wavenumber of the pattern is sufficiently far from the critical value, this instability leads to slow spatial modulations of the phase and amplitude of the pattern, the nucleation or elimination of periodic units, and eventually a new pattern with a wavenumber much closer to the critical value. This phenomenon is difficult to observe, both because it is often masked by other secondary instabilities, and because of boundary effects. We have performed the first direct observations [2] of the space and time evolution resulting from this instability, using electrohydrodynamic convection to obtain a sample containing at least 150 convection rolls (75 periodic units). By controlling the initial wavenumber of the roll pattern and the layer depth, we are able to make precise measurements of the stability boundaries and the time

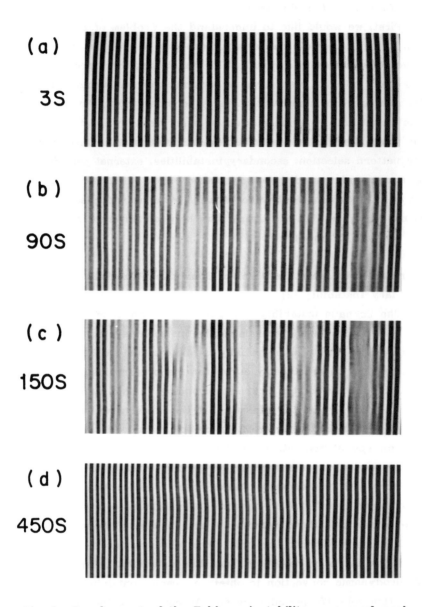

Fig. 1. Development of the Eckhaus instability, a general mechanism of pattern selection, in a spatially periodic convection pattern. New roll pairs nucleate near the minima in the convective amplitude (regions of weak contrast), leading to a higher wavenumber.

evolution of various spatial Fourier components of the pattern.

The working fluid for this experiment consists of a nematic liquid crystal (MBBA) confined between two transparent conductive electrodes with an adjustable separation d=20-120 $\mu$m. A potential difference V larger than a critical value $V_C$ (=6 volts ac) induces a one-dimensional roll pattern with critical wavenumber $k_C$ similar to that resulting from the Rayleigh-Bénard instability. We define a control parameter $\varepsilon=(V-V_C)/Vc$ and a dimensionless wavenumber $Q=(k-k_C)/k_C$.

The experiment was performed by creating a roll pattern with a fixed wavelength of 200 $\mu$m (for various d), using a small spatially periodic electric field. (The periodic field was generated with the help of a photolithographically produced interdigitated electrode.) The periodic field is then eliminated without changing $\varepsilon$, and the pattern is allowed to evolve to a steady state. If the initial conditions are located in the "Eckhaus unstable" region of the phase diagram (far from $k_C$), the initial periodic pattern develops a long wavelength modulation, as shown in Fig. 1. The amplitude of the flow approaches zero at certain locations, and new roll pairs are nucleated there. This process drives the wavenumber toward $k_C$.

The stability boundaries for the onset of convection and for the Eckhaus instability are shown in Fig. 2. The Eckhaus modulations occur only to the left of $\varepsilon_E(Q)$. If the initial conditions are chosen to lie closer to Q=0, the wavenumber still evolves toward Q=0, but by a different process involving dislocation motion. This latter process is slower by at least an order of magnitude.

The measurements of the linear stability boundary for the onset of convection, $\varepsilon_L(Q)$, are the first such measurements for a convective system. Surprisingly, this basic property is difficult to measure because periodic patterns are unstable on both sides of it. What changes when crossing this line is the mode of instability. When the pattern is linearly unstable (to the left of $\varepsilon_L(Q)$), the amplitude decays, whereas to the right of this line, the pattern is unstable with respect to amplitude and phase modulations.

This study of pattern selection near the onset of convection

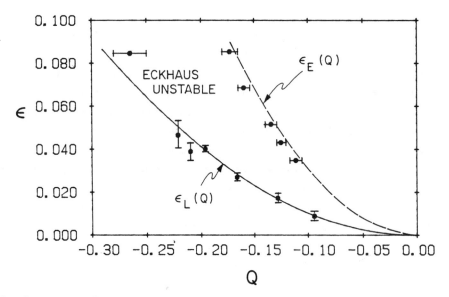

**Fig. 2.** Measured stability boundaries as a function of the distance ε above the onset of convection and the dimensionless wavenumber Q.

emphasizes the role of the Eckhaus secondary instability in determining the patterns that are realized. This mechanism is quite general, and occurs for any periodic pattern that results from a forward bifurcation. Other secondary instabilities can also affect the selected patterns.[4]

### 3. Pattern Selection by Spatial Forcing [5]

If the translational invariance of the system is broken by an external spatially periodic perturbation, the nature of the stable patterns can be dramatically changed. Pattern selection due to such spatial forcing can be conveniently studied in the context of the electrohydrodynamic convection described in Sec. 2 simply by using the interdigitated electrode to impose a steady periodic perturbation. (In Sec. 2, the perturbation was used only to establish an initial condition, and was subsequently removed.)

We found that a great variety of novel states can be produced, including: commensurate phases in which the hydrodynamic

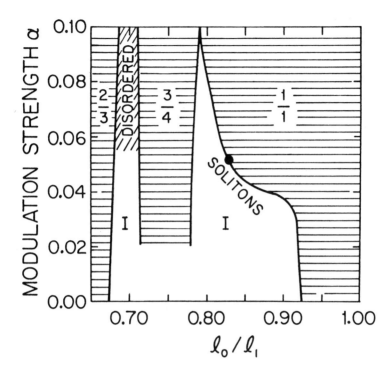

Fig. 3. Phase diagram as a function of the strength of the periodic forcing and the ratio $\ell_0/\ell_1$ of the natural hydrodynamic period to that of the forcing.

flow is "phase-locked" to the perturbation; incommensurate states containing soliton-like discommensurations that are the spatial analog of quasiperiodicity; and complex structures that contain an apparently random array of defects. These are all stable configurations of the flow in the presence of external forcing. They demonstrate the intricacy of the process of pattern selection.

Some of these phenomena can be summarized with the aid of a simple phase diagram (Fig. 3) that shows the patterns selected as a function of the strength and period of the external forcing. The regions labeled by rational numbers are phase-locked patterns in which the hydrodynamic period and the forcing are commensurate. The region marked "solitons" contains a quasiperiodic (incommensurate) spatial structure in which the rolls are periodically compressed. This allows the phase of the rolls to slip with respect to the forcing, thus leading to an incommensurate structure. The solitons have been explained recently by P. Coullet. [6] The "disordered" states noted in the phase diagram have been discussed briefly elsewhere.[7]

These states are quite different from those realized when the translational invariance of the system is not broken externally. We expect that other nonlinear systems would show similar structures in the presence of spatial forcing. These phenomena demonstrate the intricacy of the process of pattern selection near the onset of a hydrodynamic instability.

4. Pattern Selection by Boundaries and Defects [8]

In most experimental studies of Rayleigh-Bénard convection, lateral boundaries and defects in the pattern itself play a significant role in the selection of stable patterns. Most studies of these phenomena have been qualitative. We have found that some insight into the dynamics of this pattern forming process can be obtained by quantitative analysis of the structures.

An example of the evolution of a Rayleigh-Bénard convection pattern in a cylindrical cell is shown in Fig. 4. The pictures were made by digitization (and enhancement) of images formed by refraction of a parallel beam of light; the patterns show the refractive index gradients in the cell due to the inhomogeneous temperature distribution. Light regions correspond to cold decending fluid, while dark regions correspond to warm rising fluid. A disordered flow was stimulated by starting very far above the onset of convection, and then abruptly reducing the temperature gradient so that the Rayleigh number R is $2.5R_C$, where $R_C$ is the Rayleigh number at the onset of convection.

The remarkable thing about these patterns is that even after a very long time, the patterns do not resemble at all the classical form

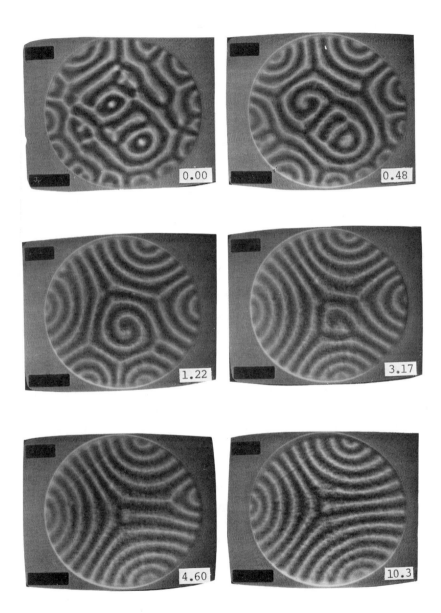

Fig. 4. Evolution of a convection pattern at Rayleigh number R=2.5R$_c$. Times are given in units of the horizontal thermal diffusion time (3 h).

predicted by stability theory: an array of stable parallel rolls. The patterns remain textured, with the rolls being oriented roughly perpendicular to the outer cell boundary. The patterns necessarily contain defects, where the local roll orientation is undefined.

There are many stable textured patterns accessible for different initial conditions; they correspond to different basins of attraction. Several examples of stable patterns that can be obtained for the same Rayleigh number are given in Fig. 5. Of course there are actually an infinite number of possible stable patterns, if rotations are taken into account. On the other hand, the range of accessible patterns is much smaller than the range given by stability theory. This is clear because the mean wavenumber is always found to be much more tightly constrained than stability considerations (for an infinite system) require.

We believe that the (circular) lateral boundary plays an essential role in selecting stable patterns in this finite system.[9] However, boundaries are not the only consideration. The expulsion of excess defects (those not required to allow perpendicular alignment at lateral boundaries) is another important factor in the dynamics.

How are we to understand this process of pattern selection? One (impractical) approach is to try to use the full hydrodynamic equations. They are simply too complex even for numerical simulations of these phenomena. The basic reason is that the phenomena are so slow that numerical integrations would be excessively time-consuming, even on a supercomputer.

The most useful approach is to use simplified model equations. We have been able to make a quantitative test of one intriguing model, due to Swift and Hohenberg [10]. The model is a two-dimensional partial differential equation for the evolution of a field that many be taken to be the temperature field in the midplane of the layer. The model has the interesting property that a functional (or pseudo-energy) of the temperature field evolves toward a (local) minimum. This pseudo-energy is composed of several parts: a boundary term that is minimized by perpendicular alignment of the

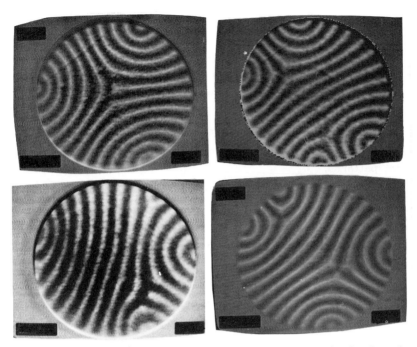

Fig. 5. Several stable convection patterns that were obtained under the same external conditions ($R=2.5R_C$). These patterns illustrate the phenomenon of multiple basins of attraction in a simple hydrodynamic system.

rolls; a defect term that is minimized by expulsion of defects; and a bulk term that is minimized by reduction of wavenumber fluctuations and curvature in the roll orientations. We have been able to test the model by digital analysis of images [8] such as those presented in Fig. 4. We do find behavior that is consistent with the model provided that R is less than about $3R_C$. (It is also necessary to exclude a small region near $R_C$ where the patterns to not stabilize.) Since minimization principles are not generally found in nonlinear systems, it is interesting that there are special situations, fairly near the onset of a hydrodynamic instablity, where a minimization principle provides an approximate description of the dynamics.

## 5. Chaos from Competing Patterns [11]

We conclude this discussion of pattern formation from instabilities by pointing out that there is an intimate connection between the problem of pattern formation and the problem of chaos. It is clear that most chaotic hydrodynamic flows involve intricate variations in spatial structure. In most cases it has not been possible to explore this relationship in detail. However, there is one simple experiment where chaos has been found to arise from the competition between two distinct patterns.

The experiment is a study of waves on the surface of a cylindrical fluid layer. The waves are driven by a small vertical oscillation of the entire container. When the driving amplitude exceeds a critical value that is a function of driving frequency, the free surface develops a pattern of standing waves. These vibrational modes are the result of a parametric instability, and are very much like the normal modes of a vibrating drum. It often happens that several modes are nearly degenerate, in the sense that their threshold curves overlap. This is demonstrated in the phase diagram of Fig. 6, which shows the various regimes as a function of driving amplitude and driving frequency. Below the parabolic stability curves, the surface is nearly quiescent. In the regions labeled ($\ell$,m), the surface is excited in the corresponding drum mode, with $\ell$ angular maxima and (m-1) nodal circles. There are many such modes, but only two are found in the narrow frequency band shown in the diagram.

The interesting·dynamics occurs in the cross-hatched regions near the intersection of the two stability curves, where energy sloshes back and forth between the two modes on a time scale much longer than that of the standing waves. The sloshing can be either periodic, or chaotic, as indicated in the diagram. In the chaotic regime, we have found that the motion can be described by a low dimensional strange attractor.[11]

In most cases where chaotic behavior has been demonstrated, the spatial characteristics of the flow have not been determined. It is fascinating to see that the chaotic motion can arise in a simple way

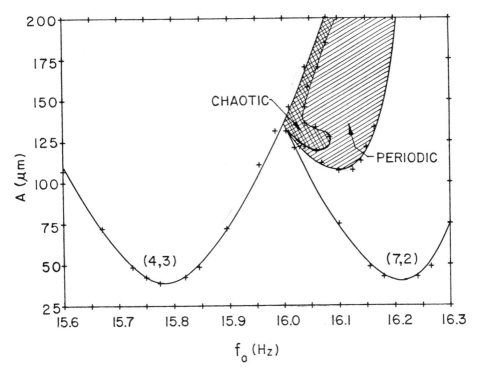

Fig. 6. Phase diagram for driven surface waves on a cylindrical fluid layer, as a function of the driving amplitude A and frequency $f_o$. Stable patterns occur near the minima of the parabolas. Slow periodic and chaotic oscillations involving the competition between these modes occur in the shaded regions.

from competition between two distinct spatial modes of the system. Thus the problem of pattern formation has a direct bearing on efforts to understand chaotic motion.

Another distinctive feature of this experiment is the fact that the chaotic motion occurs very close to the initial pattern-forming threshold. This fact should facilitate theoretical efforts to explain the observations. We have been able to account for most of the observations phenomenologically using a model of two coupled parametrically forced nonlinear oscillators. A theoretical study based on the hydrodynamic equations is in progress.[12] Preliminary results are in approximate accord with the experimental observations.

## 6. Future Developments

To a great extent, future progress depends on the development of experimental methods capable of simultaneously following the spatial structure and temporal development of hydrodynamic flows. Digital image processing methods have already demonstrated the potential to accomplish this in experimental situations where the flow is primarily two-dimensional. Three dimensional situations are more problematic. However, the difficulty is not only one of instrumentation. It is frequently not clear how to process the data. New methods of data analysis are needed that allow sensitive experimental tests of theoretical models without resorting simply to direct comparison with numerical simulations.

## 7. Acknowledgements

The work described in this review was supported by the National Science Foundation under Grant MSM-8310933. I am indebted to my collaborators, M. Lowe, M.S. Heutmaker, and S. Ciliberto. Financial support from the the John Simon Guggenheim Memorial Foundation is gratefully acknowledged.

## REFERENCES

1. For a review, see N.B. Abraham, J.P. Gollub, and H.L. Swinney, "Testing Nonlinear Dynamics," Physica 11D, 252 (1984).

2. M. Lowe and J.P. Gollub, "Pattern Selection Near the Onset of Convection: The Eckhaus Instability," to appear (1985).

3. For example, see L. Kramer and W. Zimmerman, "On the Eckhaus Instability for Spatially Periodic Patterns," Physica 16D, 221 (1985).

4. F.H. Busse, "Transition to Turbulence in Rayleigh-Benard Convection, in Hydrodynamic Instabilities and the Transition to Turbulence, ed. by H.L. Swinney and J.P. Gollub (Springer-Verlag, New York, 1985), p. 124.

5. M. Lowe and J.P. Gollub, "Solitons and the Commensurate-Incommensurate Transition in a Convecting Nematic Fluid," Phys. Rev. A 31, 3893 (1985); M. Lowe, J.P. Gollub, and T.C. Lubensky, "Commensurate and Incommensurate Structures in a Nonequilibrium system," Phys. Rev. Lett. 51, 786 (1983).

6. P. Coullet, to be published (1985).

7. J.P. Gollub and M. Lowe, "Incommensurate Phases in Convecting Liquid Crystals," Ferroelectrics, to appear (1985).

8. M. Heutmaker, P.N. Fraenkel, and J.P. Gollub, "Convection Patterns: Time Evolution of the Wavevector Field," Phys. Rev. Lett. 54, 1369 (1985); M.S. Heutmaker and J.P. Gollub, "Pattern Evolution in Rayleigh-Benard Convection, to be published.

9. Boundary effects have been discussed extensively by M.C. Cross, P.G. Daniels, P.C. Hohenberg, and E.D. Siggia, "Phase-Winding Solutions in a Finite Container above the Convective Threshold," J. Fluid Mech. 127, 155 (1983).

10. J. Swift and P.C. Hohenberg, "Hydrodynamic Fluctuations at the Convective Instability," Phys. Rev. A 15, 319 (1977).

11. S. Ciliberto and J.P. Gollub, "Chaotic Mode Competition in Parametrically Forced Surface Waves," J. Fluid Mech. 158, 381 (1985); S. Ciliberto and J.P. Gollub, "Pattern Competition Leads to Chaos," Phys. Rev. Lett. 52, 922 (1984); S. Ciliberto and J.P. Gollub, "Phenomenological Model of Chaotic Mode Competition in Surface Waves," Nuovo Cimento, to appear.

12. I. Procaccia, private communication.

# Chaos and Turbulence; is there a connection?

by

Alan C. Newell[*]

Department of Mathematics
University of Arizona
Tucson, Arizona 85721

Abstract. In this essay we discuss the relation of chaos, which is the unpredictable behavior associated with finite systems of ordinary differential and difference equations, and turbulence, which is the unpredictable behavior of solutions of infinite dimensional, nonlinear partial differential equations. The evidence that there is some connection, at least in certain regimes of parameter space, is sufficiently convincing to provide the motivation to search for analytical means for reducing the governing partial differential equations to either a finite system of ordinary differential equations or a much simpler partial differential equation of universal type. A successful reduction scheme must capture the spatial structure of the dominant modes accurately and we suggest ways of finding these structures in certain limiting situations. Five such schemes are presented and, in each case, the approximation is related in some way to the presence of a small parameter, near critical, nearly integrable or nearly periodic. One of these reductions leads to the complex Ginzburg-Landau equation, which has universal character, and its importance is stressed. In connection with this equation, we introduce the terms "wimpy" turbulence and "macho" turbulence to connote the crucial differences between the behavior of its solution in one and two space dimensions, a difference which has much in common with the contrast between two and three dimensional high Reynolds number flows because of vorticity production. In the final section, several ideas concerning the nature of high Reynolds number, fully developed turbulence are presented and the possible roles of singular solutions and "fuzzy" attractors are discussed. Throughout the essay, we argue that before much new progress is made, one has to understand the onset of spatial chaos, that is, the transition from a spatially regular state (possibly with a chaotic temporal behavior) to one in which the spatial power spectrum is broadband. This question is a major focus of our present research program.

This contribution is dedicated to the memory of Dick DiPrima, a good friend and long time colleague who left us too soon. It will also appear in the Special Proceedings of the Conference on Mathematics Applied to Fluid Mechanics and Stability dedicated to his memory.

[*]I want to thank the ONR office of Physics for providing partial support for this work.

General Discussion.

The realization that finite dimensional dissipative systems can have attractors ("strange attractors") on which the motion is everywhere unstable has brought a new perspective to nonlinear dynamics. Motion on the attractor depends sensitively on initial conditions (nearby orbits diverge locally at an exponential rate on the average) and this sensitive behavior leads to an apparently stochastic time signal with a broadband power spectrum. The finite amount of information contained in the finite accuracy specification of the initial state is eroded by the flow and once sufficient time has elapsed to uncover the unknown part of the initial data, the state of the dynamical system is unpredictable. Although Poincaré was fully aware that Hamiltonian systems could exhibit such behavior, it was not widely appreciated until the early seventies, following the pioneering work of Lorenz [1] on weather prediction models and the bold and imaginative ideas of Ruelle and Takens [2], that dissipative systems could have unstable asymptotic behavior. We will refer to this long time sensitive behavior of a finite order system of ordinary differential or difference equations as chaos. Its signatures are a broadband power spectrum and a fractal dimension.

It is of course tempting to speculate that turbulence can be explained on the grounds that the apparent stochastic time dependence of a fluid is the manifestation of sensitive dependence on a relatively low dimensional strange attractor. Appealing as this idea may sound, one must realize that a fully developed turbulent flow has a complicated spatial as well as temporal character and that any theory which purports to explain turbulence in terms of a finite set of ordinary differential or difference equations must come to grips with this fact. Chaos has broadband temporal behavior; fully developed turbulence has broadband spatial and temporal behavior. In Table 1, we list some properties of and comparisons between the two. First, however, let me make some operational definitions. Chaos is the unpredictable behavior (broadband power spectrum and at least one positive Lyapunov exponent) of systems of ordinary differential and difference equations. Turbulence is the unpredictable behavior of partial differential equations. It may have an ordered or disordered spatial structure. I use the term fully developed turbulence to refer to a flow with a disordered spatial structure (broadband spatial power spectrum and decay of spatial correlations), although when the distinction is clear from the context I sometimes refer to the latter state also as turbulence. It would probably have been more in line with present usage to refer to the former as weak turbulence; however this term is also used for the stochasticity encountered in weakly nonlinear dispersive wave systems, e.g., ocean waves.

TABLE 1

| CHAOS | TURBULENCE |
|---|---|
| Unpredictable behavior of $v = \vec{f}(\vec{v},R)$, $v \in R^m$ | Unpredictable behavior of solutions of NLPDE's |
| Dissipative | e.g. Navier-Stokes Maxwell-Bloch Envelope Equations |
| e.g. Lorenz equations Hénon and logistic maps | |
| | Space and time independent variables |
| One independent variable (time) | |
| | Potentially an infinite number of degrees of freedom |
| Finite number of degrees of freedom | |
| | Onset of turbulence |
| Broadband temporal power spectrum | Period doubling Ruelle-Takens Intermittency |
| Exponential decay of autocorrelation function | |
| | Onset of spatial disorder, gateway to fully developed turbulence |
| Sensitive dependence strange attractors | |
| Fractal dimension | Broadband spatial and temporal power spectra |
| Positive Lyapunov exponents | |
| | Exponential decay of time and space correlations |
| Universal (self similar) behavior | |
| Routes to chaos | High levels of fluctuation vorticity and rapid stretching of vortex filaments |
| Period doubling Ruelle-Takens Intermittent behavior | |
| | Diffusivity and rapid mixing |
| | Dynamic similarity, self similar ranges of spectrum |
| | Statistical closure problem |

To understand turbulence, one must have a means of identifying the spatial structures which dominate the flow field. In fully developed turbulent flow, these are obviously very complicated. Even near the onset of the broadband spatial spectrum, just past that value of the stress parameter at which the spatial structure is relatively coherent, there is marked patchiness and intermittency. A good example of this is the sudden appearance of streaks and bursts in boundary layers which can be seen a few boundary layer thicknesses downstream of the onset of Tollmien-Schlicting waves. A natural question arises; is it possible to find a finite dimensional basis which uniformly captures all this erratic behavior? To be sure, there are recent theorems by Ruelle [3], Foias, Temam, Manley, Trève [4], Nicolaenko, Hyman [5], and others which place an upper bound on the dimension of attractors of dissipative partial differential equations, but, while their existence is comforting, these upper bounds are very large and in all likelihood not of any great practical use. For example, Ruelle obtains an upper bound on the Hausdorff dimension of three dimensional Navier-Stokes flow in terms of the viscous dissipation rate $\varepsilon$ (assuming global solutions exist), but this bound is equivalent to the intuitive notion that it is sufficient to resolve the flow field down to the viscous dissipation scale (Kolmogoroff's inner scale) $(\nu^3/\varepsilon)^{1/4}$. This would require $\ell^3/(\nu^3/\varepsilon)^{3/4}$ Fourier modes, which is proportional to the nine-fourths power of the Reynolds number. ($\varepsilon$, the viscous dissipation rate can be written as $U^3\ell^{-1}$ where $U$ is a typical large eddy velocity and $\ell$ is a large eddy length scale). This means that for Reynolds numbers of $10^4$, one would need a billion modes to resolve the system. It is hardly likely that replacing the Navier-Stokes equations by a system of a billion ordinary differential equations for the Fourier mode amplitudes will lead to much simplification or an increased understanding of turbulent processes. Consequently, it is fairly clear that, for large Reynolds number flows, some additional ideas will be needed.

Nevertheless the evidence which supports a link between real, potentially infinite dimensional, flows in certain limited parameter ranges and finite dimensional mathematical models, while indirect, is both convincing and compelling. Before we review the evidence, let us briefly mention the two categories of parameter which directly influence dimension. The first category includes those stress parameters such as Reynolds, Rayleigh, Chandraskkar, Taylor numbers, loading stress, which measure the external force applied to the system. The second parameter of importance is the aspect ratio, which is the ratio of the largest container dimension to a natural length introduced by the dynamics, and measures the constraint which the geometry has on the flow. Other parameters, such as the Prandtl number, which are ratios of internal competitive influences, are of course important but do not usually determine to any large extent the effective number of active degrees of freedom (by affecting stability characteristics, a change in these parameters may change the dimension by a small amount).

Large values of the stress parameter  R  and aspect ratio  Γ
potentially allow the flow to explore a greater number of its degrees
of freedom and a greater phase space volume.  Low to moderate values
constrain the allowed motions.  The phenomenon of fully developed
turbulence is associated with large values of the stress parameter (in
pipe flow the Reynolds number may be $10^3$; in the sun the Rayleigh
number can be $10^7$) in open (large aspect ratio) systems.  On the other
hand, the values of the stress and aspect ratio parameters in which
links are found between real experiments and mathematical models are
in the low to moderate ranges and usually involve regimes of
transition to temporally chaotic but spatially ordered behavior.

The way in which these links between real or numerical experiment
and the various finite dimensional mathematical models (the universal
logistic and circle maps, for example) are established is through an
analysis of the time record of some measured or computed quantity.
The reliability of such analyses has depended heavily on the greatly
improved methods of accurate measurement (e.g., laser Doppler) together
with the data acquisition and processing power of modern computers.
Furthermore, one has to have great confidence in the long time control
of boundary conditions and sources of external noise because it is
crucial to distinguish between the apparent stochasticity introduced
by the dynamics and that due to external noise.  Despite the
experimental difficulties, however, such control, while requiring
considerable ingenuity on the part of the experimentalist, is
possible.  Several of the leaders in this most important endeavor have
been Ahlers, Bergé, Gollub, Libchaber, and Swinney.  There are two ways
in which the time signal is analyzed.  First, one constructs its
spectral content by calculating the Fourier transform of its
autocorrelation function.  A spectrum which consists of discrete
spikes corresponds to a quasiperiodic flow on a torus.  A single spike
and its harmonics means the flow is periodic, two fundamental
incommensurate frequencies with all integer combinations of sums
corresponds to a quasiperiodic flow on a two-torus.  Since the
fundamental frequencies are functions of the motion amplitude, it
turns out that for certain ranges of the stress parameters, they may
fall into a locked state in which the motion is again periodic and the
torus is not densely covered by the solution orbit.  Quasiperiodic
flow on a three torus (three incommensurate frequencies) is possible
but rare, the exception rather than the rule, and destabilizes to a
nearby strange attractor.  This is the Ruelle-Takens-Newhouse [6]
result, an update on the original idea of Ruelle and Takens [2] who
proposed the same fate for a four torus.  (A word of warning; physics,
because of sometimes obvious and moretimes subtle symmetry
constraints, is not always generic).  The emergence of a broadband
component in the spectrum signals the onset of sensitive motion on the
strange attractor but it could also simply be due to external noise.
In order to distinguish between these two possibilities, one employs a
second means of data analysis, reconstructing an m-dimensional phase
portrait of the flow by using as coordinates  m  dimensional subsets
of the time series itself.  This construction, justified by an

application of the Whitney imbedding theorem by Takens and Mañé, can
be used to measure both the dimension (capacity, pointwise or
correlation dimension, see [7]) and the Lyapunov exponents (at least
the largest positive ones) of the flow. If the broadband behavior is
due to noise, the measured dimension will increase with imbedding
dimension; if due to a strange attractor, it will converge once the
imbedding dimension is large enough. Roughly speaking, m, the
imbedding dimension, should be 2d+1 where d is the next greatest
integer to the Hausdorff dimension of the attractor.

By this kind of analysis, one can monitor in a quantitative way the
changes that occur in the power spectrum, the dimension and Lyapunov
exponents as the stress and aspect ratio parameters are varied. In
particular one can observe those regimes just before, during and after
the emergence of a broadband spectral component. What is truly
remarkable is that the analysis on real fluid experiments (e.g.
convection in low aspect ratio boxes [8]), in real situations in
bistable optical cavities [9] has revealed that the onset of chaotic
behavior occurs in a manner very similar to that predicted by certain
universal scenarios which arise in the investigation of simpler
mathematical models. Specifically, Libchaber, Favor, and Laroche [8]
clearly observe both the period doubling and the Ruelle-Takens routes
to chaos in carefully controlled convection experiments. Further,
Fein, Heutmaker, and Gollub [10], and Libchaber [11] confirm that the
breakdown of two torus flow follows very closely the predictions of
the circle map [12,13] which can plausibly be argued to capture the
universal features of the transition in this case. Not only are the
results in qualitative agreement, but the various universal numbers
(convergence rate of the period doubling sequence, the relative
amplitudes of the spectral harmonics, the geometric convergence rate
of the winding numbers in the circle map), emerge from the
experimental data. The reader may wish to refer to references [14,15]
for more details of comparison.

It is important to realize, however that the link is not deductive.
A deductive link would involve the development of a rational, finite
dimensional approximation to the nonlinear partial differential
equations in question, an approximation which is asymptotically valid
under certain limiting conditions. Having developed the approximation
to this stage, one can then refine it further, by the use of Poincaré
maps or perhaps using numerical results as a guide to argue its
relation with one of the by now standard mathematical models which
capture the various transition scenarios. This process of building
simplified rational approximations to more complicated equations is
called model building. The acid test comes when one directly compares
the predictions of the reduced model with the results of the real
experiment. On the other hand, what is presently taken to be a point
of contact between experiment and the behaviors exhibited by certain
"model" equations is more accurately described by the nouns metaphor
or simile [16] (a figure of speech in which one thing is likened to
another). While the experiment of Libchaber et al [8] strongly
suggests that a direct deductive connection between the Oberbeck-

Boussinesq equations and the logistic map may be found, it does not in itself prove this connection. Just because it walks like a duck, quacks like a duck, looks like a duck doesn't mean it is one! Nevertheless the evidence provides a powerful motivation for finding direct links. It is to ways of accomplishing this task that this essay is addressed.

The key step in the challenge is to identify a priori the spatial structures (the active modes) which dominate the dynamics. In short, one would like to identify the correct finite dimensional basis into which to project the field variable, a basis which is uniformly valid over long time; that is, no new modes are produced in any substantial way by the dynamics. In certain cases, this is a simple task, a straightforward application of the center manifold theorem. In others, where the flow is fully nonlinear but still spatially coherent, one can look for multiparameter, fully nonlinear asymptotic states; even here a fair amount of guesswork and physical intuition is involved. Finally, when the spatial coherence is lost and spatial correlations decay, one has to deal with very complicated spatial behavior. This may be a manifestation of the sudden activation of a large number of modes or it may be the development of very complicated, but still low dimensional, fractal spatial structures. We will return to a discussion of this point at the end of the essay.

Our goal is to complete the right hand side of the diagram in Figure 1. It would also be of value to understand how one might construct the map $\Phi$ between the coordinates constructed from the signal and the natural coordinates which parametrize the active modes, perhaps using several reconstructions from different parts of the signal data. In other words, it would be very useful if we could somehow gain some information on the choice of natural coordinates from the time series itself.

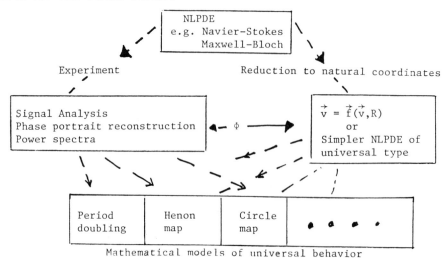

Figure 1. Reduction to finite dimensional systems.

Reduction of the governing equations.

There are six limiting cases which fall into three categories; near transition, moderate and large Reynolds number. By the transition Reynolds number, we mean that value $R_c$ of the stress parameter at which the stability of the stationary solution to the partial differential equation is lost. Often, this solution is a global fixed point attractor, for example, the purely conductive state in the Rayleigh-Bénard problem.

Category I. Near transition $R \sim R_c$

Ia. Low to moderate aspect ratio
Ib. Large aspect ratio

Category II. Moderate values of $R$

IIa. The Galerkin truncation
IIb. Nearly integrable systems
IIc. Nearly periodic systems

Category III. Large values of $R$

Our goal in each case is to find a rational approximation to the full nonlinear partial differential equation. In some cases (e.g., Ia), this will consist of a finite set of ordinary differential equations; in others, it will consist of a very much simpler partial differential equation of universal type. Specfically in IIa, we will introduce the complex Ginzburg-Landau equation which describes the evolution of the envelope of the most unstable wave when the stress parameter $R$ reaches its critical value. This equation has some very interesting properties not the least of which is the constrast of its behaviors in one and two space dimensions. Indeed, its long time solution has many features analogous to the Navier-Stokes equations in two and three dimensions respectively.

In all cases, we seek to take advantage of the smallness or largeness of some parameter which will serve as the perturbation parameter in an asymptotic expansion for the field variable of the partial differential equation. In category I, this small parameter is proportional to $R - R_c$ and the amplitude of the resulting flow field. In category II, one can no longer take advantage of small amplitudes and weakly nonlinear flows and in most cases one loses the advantage of using a linear superposition of the active modes as a first approximation to the flow field. Therefore, in this category, one must take advantage of some other special quality of the system. In particular, we will use the fact that in many nonlinear optical situations, the electric field envelope satisfies an equation of nonlinear Schrödinger type. In such cases, one then can attempt to describe the dynamics in terms of the fully nonlinear asymptotic states of the exactly solvable nonlinear Schrödinger equation.

As I have mentioned, the key step in obtaining a rational approximation to the flow field $\vec{u}(\vec{x},t)$ which satisfies the original equations (the Navier-Stokes, Maxwell-Bloch equations, etc.) is to identify correctly and a priori the spatial structure of the modes which take an active part in the dynamics. In weakly nonlinear systems, this structure is provided by the "unstable" or "neutral" eigenfunctions of the linear operator which investigates the stability of the solution whose stability is about to be lost. These structures then serve as basis vectors of the subspace into which we project the field variable $\vec{u}(\vec{x},t)$. The amplitudes (or envelopes) of the unstable eigenfunctions are the coordinates and we can attempt to follow their time (or time and space) evolution. Providing that the degeneracy problem can be handled (see upcoming discussion in Ia,b), one can have confidence that the field variable $\vec{u}$ is uniformly well approximated by its projection into this subspace.

For moderate values of the stress parameter for which the field amplitudes are not small, one has the problem of identifying the spatial structure of the solution branches. In principle, one can always follow the attracting solution branch after the first bifurcation from the simplest state at $R_c$ and continually monitor its local stability properties. However, it is very difficult to do this in practice, and particularly so when the system jumps from one solution branch to another at the coalescence of unstable saddles and stable nodes. For example, in IIb where we look at the dynamics of optically bistable ring cavities, it turns out that when the amplitude of the central part of the input beam exceeds a certain magnitude, a whole new set of spatial structures (which correspond to fully nonlinear solitary wave solutions of the underlying propagation equation) are formed.

As an alternative means of analysis, one can take adventage of the existence of nearby exactly integrable systems when such neighbors exist. We will argue later on that the universality of equations of nonlinear Schrodinger type and particularly their relevance in nonlinear optics makes this a viable alternative particularly when one is dealing with the Maxwell-Bloch equations. The idea is to analyze the behavior of the system of interest by projecting the field variable into a subspace which consists of a multiparameter, exact nonlinear solution of its integrable neighbor,

$$\vec{u}(\vec{x},t) = \vec{F}(\vec{x},t; A_1, A_2, \ldots, A_N). \tag{1}$$

For example, $\vec{F}$ could be a two soliton solution of the nonlinear Schrodinger equation and the soliton amplitudes and positions would serve as coordinates in the subspace in which one believes the important dynamics occurs. Because the real equation is only close to the integrable equation, the time evolution of these coordinates, which can be found by perturbation theory, is nontrivial. I will outline some of the success in exploiting these ideas in IIb. In relatively simpler situations in which one can guess a priori what the active modes will be, we have had considerable success. One can even get away with perturbing around systems which are not exactly integrable but which do possess fully nonlinear, isolated, stable solutions. If the dynamics involves such isolated structures, one can use a "linear" superposition of these modes because in the overlap regions the pulses are exponentially small. Despite this limited success, however, it is fair to say that the general problem of identifying the active modes, that is the subspace into which to project the field variable, remains wide open.

Ia. <u>Near transition. Low aspect ratio</u>. In this category, the spectrum of the linearized equations describing the variation of the solution about to lose stability is discrete and at most finitely degenerate. Thus, as R crosses $R_c$, only a finite number m of

eigenvalues have real parts which lie on or cross over the imaginary axis. The eigenvalues (which may be complex) measure the growth rates and oscillatory behavior of the eigenfunctions $\{\phi_n(\vec{x},t)\}_1^n$ of the

linearized equation. Near $R_c$, we can approximate the field $\vec{u}(\vec{x},t)$

to leading order by a linear superposition of these eigenfunctions (the center manifold theorem)

$$u(x,t) = \varepsilon \sum_{n=1}^m A_n(T = (R-R_c)t)\phi_n(\vec{x},t) + \varepsilon^2\vec{u}_1 + \ldots , \qquad (2)$$

where $\varepsilon$, the amplitude measure is related to some power of $R - R_c$. By demanding that (2) is a uniformly valid asymptotic expansion for solutions $\vec{u}(\vec{x},t)$ of the <u>nonlinear</u> variational equation, we can develop a system of ordinary differential equations for the mode amplitudes $\{A_n(T)\}_1^m$,

$$\frac{dA_n}{dt} = f_n(A_s), \quad n,s = 1,\ldots,m. \qquad (3)$$

A simple example is

$$\frac{\partial w}{\partial t} + (\nabla^2+1)^2 w - Rw + w^3 = 0 \qquad (4)$$

on the square $0 \leq x,y \leq L$ with zero boundary conditions on $w(x,y,t)$ and its second derivatives. The global attractor $w = 0$ loses its stability at $R = 0$ and a linear stability analysis gives

$$\sigma_{mn} = R\left(-\frac{(m^2+n^2)\pi^2}{L^2} - 1\right)^2 , \qquad (5)$$

where

$$w(x,y,t) = 0 + \sum A_{mn} \sin \frac{m\pi x}{L} \sin \frac{n\pi y}{L} e^{\sigma_{mn}t} .$$

The neutral curves $\sigma_{mn} = 0$ are shown in Figure 2.

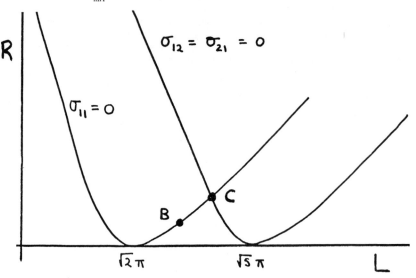

Figure 2. Neutral stability curves in R,L plane.

For parameter values near B, there is only one active mode $\sin \frac{\pi x}{L}$ $\sin \frac{\pi y}{L}$ and its amplitude satisfies the Stuart-Watson [17] equation

$$\frac{dA}{dt} = (R-R_B)A - \frac{9}{16} A^3 , \qquad (6)$$

whose attractor is a stable fixed point. For parameter values near C, three modes $m = n = 1$, $m = 1$, $n = 2$ and $m = 2$, $n = 1$ are active.

These kind of amplitude equations have been analyzed in great detail in many contexts. Reference are given in [18]. In most cases to date, the finite dimensional system (3) has only fixed point or limit cycle attractors. However for $m \geq 3$, there is no reason in general to exclude strange attractors. For example, Busse has found a number of finite dimensional situations [19] in which one obtains chaotic behavior immediately after $R = R_c$. In one of these, convection in a horizontal layer with stress free boundaries, there are no stable convective rolls as the oscillatory skewed varicose and skewed varicose instability boundaries in the Busse balloon become coincident at a Prandtl number value of 0.543.

In summary, then, for low aspect ratio flows near $R_c$, the spatial structure of the active modes can be determined and the behavior of the solution of the original field equation is well described by the finite order system of ordinary differential equations for the mode amplitudes.

Ib. Near transition. Large aspect ratio. There are two kinds of infinite degeneracy which occur in systems where the boundaries are at infinity or too far away to constrain the spectrum of spatial modes. The first is an exact degeneracy and reflects a continuous symmetry. For example, if we let $L \to \infty$ in (6) and if

$$w(x,y,t) = 0 + \sum_{\vec{k}} A(\vec{k}) \, e^{i\vec{k}\cdot\vec{x} + \sigma t} , \tag{7}$$

then the growth rate

$$\sigma = R - (k^2-1)^2, \quad k = |\vec{k}| = \sqrt{k_x^2 + k_y^2} . \tag{8}$$

Clearly once $R > 0$, all wavevectors $\vec{k}$ lying on the circle $|\vec{k}| = 1$ grow at the same rate. The orientational degeneracy reflects the rotational symmetry of the Laplacian. This degeneracy is the source of the weak turbulence seen in convective patterns [20] in large aspect ratio boxes as rolls of different orientation endlessly compete for dominance. We return to a discussion of this case in IIc.

For now, we will be more interested in the second type of degeneracy, and "almost" degeneracy which arises when the spectrum is (or almost is) continuous. Observe from Figures 3a,b that if $R > R_c$, (in example (6), $R_c = 0$), all wave numbers in the band AB of Fig. 3a and all wavevectors in the shaded annulus of Figure 3b are excited.

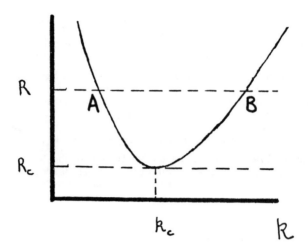

Figure 3a:   Neutral curve   R vs k

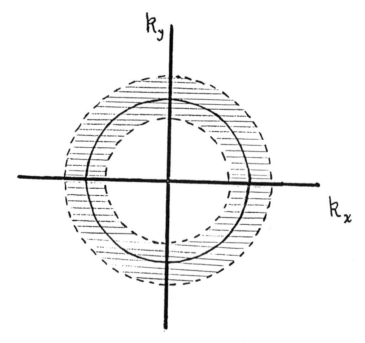

Figure 3b:   Annulus of excited modes

How are all these included in the nonlinear analysis? In IIc, I will discuss how the orientational degeneracy is usually treated. Here we will focus on the second type of degeneracy and in order to remove the orientational degeneracy from the picture, we will assume that some external influence breaks the rotational symmetry and that a unique wavevector $\vec{k}_c$ is preferred. From the linear stability diagram, Figure 3a, it is clear that a continuous spectrum is excited when $R > R_c$. While there is no rigorous analogue to the center manifold theorem in this case, the finite bandwidth effects can be captured by introducing a slowly varying envelope function to modulate the shape of the most unstable and unique eigenmode. We are also going to assume that the most unstable mode $\vec{k}_c$ is overstable; that is, the instability sets in as a growing wave (unlike the case in example (4)). Specfically if $\vec{x}$ is the vector of directions of infinite extent and $\vec{z}$ the directions in which the spectrum is discretized, we set the field variable

$$\vec{u}(\vec{x},\vec{z},t) = A(\vec{X} = \epsilon(x - \nabla\omega t),\ T = \epsilon^2 t)\ \vec{\phi}(\vec{z})e^{i(\vec{k}_i \cdot \vec{x} - \omega_c t)} + (*)$$
$$+ \epsilon\vec{u}_1 + \epsilon^2\vec{u}_2 + \dots, \tag{9}$$

where $\omega = \omega(\vec{k},R)$ is the dispersion relation of the growing wave, $\nabla\omega$ is the group velocity of the most critical wave and $\vec{\phi}(\vec{z})$ captures the eigenfunction structure in the directions of finite extent. We have defined $\epsilon$ to be $\sqrt{R - R_c}$.

The envelope function $A(\vec{X},T)$ satisfies a universal equation

$$\frac{\partial A}{\partial T} - \frac{1}{2} \sum (\frac{\partial\sigma}{\partial R} \frac{\partial^2 R}{\partial k_j \partial k_\ell} + i \frac{\partial^2\omega}{\partial k_j \partial k_\ell}) \frac{\partial^2 A}{\partial X_j \partial X_\ell}$$

$$= (\frac{\partial\sigma}{\partial R} - i \frac{\partial\omega}{\partial R})A - (\beta_r + i\beta_i)A^2 A^*. \tag{10}$$

All the coefficients of the linear terms are obtained from the linear stability analysis which provides the growth rate $\sigma(\vec{k},R)$ and frequency $\omega(\vec{k},R)$ as functions of $\vec{k}$ and $R$. The neutral stability curve is $\sigma = 0$, giving $R = R(\vec{k})$, and the lowest value of $R$ on this curve occurs at $\vec{k} = \vec{k}_c$ when $R = R_c$. Thus the positive diffusion matrix $(\gamma)^r = \frac{1}{2} \frac{\partial\sigma}{\partial R} \cdot \frac{\partial^2 R}{\partial k_j \partial k_\ell}$ is the product of the growth rate of the most critical wave and the curvature of the neutral curve in Figure 3a at $\vec{k}_c$ . It measures how fast the sidebands grow relative to the most unstable wavevector. The imaginary coefficient matrix $(\gamma)^i = \frac{1}{2} \frac{\partial^2 \omega}{\partial k_j \partial k_\ell}$ measures wave dispersion. The imaginary part of the linear growth term in (10) can be removed by taking the frequency $\omega(\vec{k}_c, R_c)$ in the exponent of (9) to be estimated at $R$ rather than $R_c$. The "Landau" constants $\beta_r$, $\beta_i$ measure the nonlinear response and have the same values as would be computed from a Stuart-Watson analysis in which no sidebands are included. The real part $\beta_r$, if positive, helps the growing wave saturate; the imaginary part measures the nonlinear modulation of the frequency. Sometimes, the nonlinear terms are a bit more complicated and must include contributions from a "mean" component $B$ which in turn is related to gradients of $AA^*$ by an auxiliary equation.

Equation (10), sometimes called the complex Ginzburg-Landau equation, was originally derived by us [21] as the dispersive wave analogue of the Newell-Whitehead-Segel equation [22] which obtains when the most unstable mode is not oscillatory and the principle of exchange of stabilities holds. It is universal in character and is canonical in the sense that near $R = R_c$, the shape of the equation is problem independent (except for the possibility of a separate "mean" component). Furthermore, it has extremely rich dynamics. If $\beta_i$ and $(\gamma)^i$ are zero, then the equally preferred solutions $A = \pm \sqrt{\chi/\beta_r}$ , $\chi = \partial\sigma/\partial R$ are attractors. The solution attempts to become $X$ independent (locally the preferred wavevector $\vec{k}_c$ dominates), and the intermediate stages of the solution development involve regions of different signed amplitudes, $\pm \sqrt{\chi/\beta_r}$, which are separated by domain walls. These slowly disappear, leaving one of the attractors as the dominant one. On the other hand, if $\beta_i$, $(\gamma)^i$ are nonzero, a whole new scenario can occur. Specifically if the matrix

$$\beta_i(\gamma)^i + \beta_r(\gamma)^r \tag{11}$$

is non positive, the $\vec{X}$ independent solution

$$A = \sqrt{\chi/\beta_r} \, \exp\{- \frac{i\beta_i \chi}{\beta_r} T\} \qquad (12)$$

corresponding to dominance by the most excited wave $\vec{k}_c$, is unstable [23]. This instability is closely related to the Benjamin-Feir or modulational instability of the nonlinear Schrodinger equation (equation (10) with all real coefficients except that of $\frac{\partial A}{\partial T}$ equal to zero)

$$\frac{\partial A}{\partial T} - i \sum_{j,\ell} \gamma_{j\ell}^i \frac{\partial^2 A}{\partial X_j \partial X_\ell} = - i\beta_i A^2 A^*. \qquad (13)$$

In one space dimension, for which $(\gamma)^i$ and $(\gamma)^r$ are scalars, the envelope develops into a sequence of solitary wave like states [21, 2nd reference] and a weak or phase ("wimpy"!) turbulence occurs in which one loses exact predictability of the positions of these lumps. Such behavior is reminiscent of the kind of weak turbulence one observes in a field of distributed vorticity in the two dimensional Euler equations.

In two dimensions, however, if $\beta_i(\gamma)^i + \beta_r(\gamma)^r$ is negative, things are much more dramatic. In the limit of zero growth rate $\chi$, (with $\chi/\beta_r$ finite) and zero diffusion (think of this as the "high Reynolds number" limit), the corresponding nonlinear Schrodinger equation (13) develops finite time singularities. These explosive filaments are well known in nonlinear optics and are believed to be very important participants in Langmuir turbulence (the collapse states [24]) although in that case, one does have to adjoin a separate equation for the mean component (The equation pair are known as the Zakharov equations). In these circumstances, the complex Ginzburg-Landau equation behaves like the high Reynolds number limit of the three dimensional Navier-Stokes equations in that the filamentation is analogous to the rapid stretching of vortex filaments by the "Euler" component of the Navier-Stokes equations.

What is remarkable is that the exact solutions, the solitary waves in the one dimensional complex Ginzburg-Landau equation and the self-similar filaments in the two dimensional one are the spatial structures which appear to dominate the long time dynamics even in the strongly turbulent regime. This observation gives further credence to our notion that turbulence is dominated by fully nonlinear exact solutions of the governing field equations which are asymptotically robust. This does not mean that they exist for all time; they can have finite time singularities. What is does mean is that they emerge from the flow field as coherent eddies which become isolated by dispersing away that part of their energy which does not conform to their asymptotic shapes. Eventually when they develop small scale structure, other mechanisms such as Landau damping, viscosity, diffusion, which are initially neglected, become important and lead to their destruction.

II. <u>Moderate values of R</u>. The recipes which fall into the second or global category and obtain when $R - R_c = O(1)$ cannot take advantage of a small amplitude expansion. Instead they have to rely on another small parameter ("nearly" integrable, "almost" periodic etc.) or else make some severe ad hoc assumptions not justified by any rational hypothesis. We will now discuss three.

IIa. <u>The Galerkin projection</u>. In this method, one projects the field variable $\vec{u}$ into a Galerkin basis

$$\vec{u} = \sum_{j=1}^{N} A_j(t) \, \vec{\phi}_j(\vec{x}) \tag{14}$$

consisting of a finite number $N$ of basis vectors belonging to a complete set. The basis elements are usually chosen to satisfy boundary conditions. Because the sum of such modes is not an exact solution, nonlinear products must be re-expanded and only that part of the re-expansion which falls into the subspace spanned by the finite basis is retained. By equating coefficients of the basis vectors, a system of $N$ ordinary differential equations for the $N$ mode amplitudes $A_j(t)$, $j = 1,\ldots, N$ is found. As I have mentioned, there are recent theorems which suggest that the projection gives a uniformly good approximation to $\vec{u}$ for the full Navier Stokes equations where $N$ is the number of Fourier modes required to resolve the volume down to the viscous dissipation scales and is proportional to the nine fourths power of the Reynolds number. However if $N$ is chosen too small so that the basis $\{\phi_j(\vec{x})\}_1^N$ does not accurately capture the spatial structure of the active modes, spurious results can be found.

A good example of a set of equations derived in this manner is the Lorenz equations [1]. Let $\vec{u} = \nabla \times \psi \hat{y}$ be the two dimensional velocity field of a fluid in an infinite horizontal layer of depth $d$. Then the Oberbeck-Boussinesq equations for the scalar vorticity $- \nabla^2 \psi$ and convective temperature field $\Theta$ are

$$\frac{\partial}{\partial t} \nabla^2 \psi = \frac{\partial(\psi, \nabla^2 \psi)}{\partial(x,z)} + \nu \nabla^4 \psi + \alpha g \frac{\partial \Theta}{\partial x} , \tag{15a}$$

$$\frac{\partial}{\partial t} \Theta = - \frac{\partial(\psi, \Theta)}{\partial(x,z)} + \beta \frac{\partial \psi}{\partial x} + k \nabla^2 \Theta. \tag{15b}$$

Let us expand the fields $\psi, \Theta$ in a truncated version of the basis

$$\{(\sin \frac{mx\pi a}{d}, \cos \frac{mx\pi a}{d}), \sin \frac{n\pi z}{d}\} \tag{16}$$

as follows:

$$\psi = \frac{(1+a^2)k}{a} \sqrt{2} \, X(t) \sin \frac{\pi x a}{d} \sin \frac{\pi z}{d} , \tag{17a}$$

$$\Theta = \frac{\beta d R_c}{\pi R_a} (\sqrt{2} \, Y(t) \cos \frac{\pi x a}{d} \sin \frac{\pi z}{d} - Z(t) \sin \frac{2\pi z}{d}). \tag{17b}$$

Here $\frac{2\pi d}{a}$ is a horizontal scale of the convection cell, $R_a = \frac{\alpha g \beta d^4}{\nu k}$

is the Rayleigh number, and $R_c = \frac{(1+a^2)^3}{a^2} \pi^4$ its critical value. The

Jacobian in (15a) is identically zero. The Jacobian in (15b) has only one term which lies outside the basis. It is proportional to $XZ \cos \frac{\pi x a}{d} \sin \frac{3\pi z}{d}$. We ignore it. Matching the other coefficients, we

obtain the celebrated Lorenz equations $(\tau = \frac{\pi^2(1+a^2)}{\alpha^2} \kappa t, \; b = \frac{4}{1+a^2},$

$p = \frac{\nu}{\kappa}, \; r = \frac{R_a}{R_c} )$

$$\frac{dX}{d\tau} = - pX + pY$$

$$\frac{dY}{d\tau} = rX - Y - XZ \tag{18}$$

$$\frac{dZ}{d\tau} = - bZ + XY.$$

They are only a valid approximation to the original equations when r is close to or less than unity, the value at which the conductive solution breaks down. Near that value, they are completely equivalent to the Stuart-Watson equation obtained by the method outlined in Ia for the onset of convection. They do however give a qualitatively similar picture to the real flow field for r ~ 10-20 but fail to capture the intensification of the upflux (mainly due to a neglect of the higher harmonics) to narrower and narrower regions.

Sadly, but not suprisingly, it turns out that as one uses progressively more modes N in the basis, the parameter to which the dynamics appears to be most sensitive is N, the truncation level. Not a very satisfactory state of affairs!

Such a truncation scheme has a chance of working only when the basis functions accurately describe the spatial structure in the finite amplitude $(R - R_c = O(1))$ region. For example, this is the case for convection of a Boussinesq fluid in a vertical circular annulus. Here Howard and Malkus [25] have shown that the flow components in the higher harmonics in the azimuthal angle $\phi$ $(\cos n\phi,$ $\sin n\phi$, $n \geq 2)$ decay exponentially in time and the Lorenz equations give an exact long time description of the dynamics providing that averaging over the annulus cross-section is legitimate. However, this case is very special and in general it is difficult to find an appropriate basis. If one uses an innapropriate basis, one will need a large number of modes to obtain a convergent description and picking the low dimensionality out of the large system of amplitude equations may not be easy.

What one really needs is a nonlinear basis which contains the natural asymptotic states of the partial differential equation under investigation. However, multiparameter families of exact solutions of nonlinear partial differential equations are extremely rare. There is, however, a class of such equations, the soliton equations, for which general solutions can be found and which can be written as a "nonlinear" superposition of the soliton and phonon states. A soliton is represented by a single point in the spectrum of an appropriately chosen operator. In a Fourier basis (the inappropriate basis), it would take an infinite number of modes to approximate the same structure. Moreover, one can study the deformation of the spectrum by perturbations. It is to these systems that we next turn our attention.

II(b). Near Integrable Systems. In the last fifteen years, the soliton revolution has uncovered a vast list of integrable partial differential equations many of which have direct relevance particularly in one dimensional physics. Although not yet proved rigorously, these systems have many of the same qualitative properties as their finite dimensional counterparts. Phase space is foliated by nested tori and the particular torus on which the motion takes place is determined by the constants of the motion, the action variables. The angle variables are coordinates on the tori. It is fairly natural, then, to imagine that the effects of external influences (forcing, dampling, coupling to other systems) will be similar to the finite dimensional case, and that the perturbed phase space will consist of a mosaic of islands of integrability and areas of stochasticity. For weak external influences and away from resonances, the goal is to capture the long time behavior of the perturbed system by following the dynamics of the parameters of the unperturbed solution, namely the action variables which identify the active normal modes. We describe two situations.

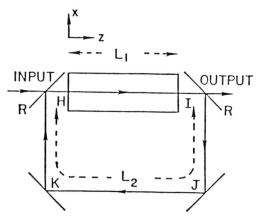

Figure 4. Ring cavity configuration

First, consider a laser beam propagating in a ring cavity containing a material with a nonlinear index of refraction. The goemetry is shown in Figure 4. The situation is described by the Maxwell-Bloch equations, and if the cavity round trip time is much larger than the relaxation times for the nonlinear medium, the equation for the envelope $B_n(\vec{x},z)$ of the electric field in the medium is

$$2iB_{nz} + \gamma\nabla^2 B_n + \beta N(B_n B_n^*)B_n = 0. \tag{19}$$

In (19), $z$ is distance along the cavity, $\vec{x}$ the transverse direction, $N$ the refractive index, $\gamma$ is the inverse Fresnel number, $\beta$ essentially measures the strength of the nonlinearity. The dependent variable $B_n(\vec{x},z)$ is the electric field envelope (the carrier is $\exp i(kz - \omega t)$) on the $n^{th}$ pass through the medium, namely the envelope pulse which begins at $z = 0$ during the times $(n-1)L/c$ $< t < nL/c$ and ends at $z = L = L_1 + L_2$ during the next time interval $(\frac{nL}{c}, \frac{(n+1)L}{c})$. The influences of the external input field $A(\vec{x}) \exp i(kz - \omega t)$ and the damping by the mirrors I and H is felt through the boundary condition

$$B_{n+1}(\vec{x},0) = A(\vec{x}) + Re^{i\phi}B_n(\vec{x},L_1) , \tag{20}$$

where $z = L_1$ is the end of the nonlinear medium, $R < 1$ is the reflectivity of the mirrors and $\phi$ is a detuning phase shift. The input envelope $A(\vec{x})$ often has a Gaussian like shape. Equation (20) is an infinite dimensional map for the envelope function $B_n(\vec{x},0)$ in which $B_n(\vec{x},L_1)$ is found in terms of $B_n(\vec{x},0)$ by solving (19) with initial condition $B_n(\vec{x},0)$. Our goal is to find the long time behavior of the envelope function $B_n(\vec{x},L_1)$ as $n \to \infty$. What we find is that the attractors for the map (20) consist of the readily identifiable asymptotic states of the nonlinear Schrödinger-like propagation equation (19). Solitons (for the case $N = 2I$, called the Kerr nonlinearity, whence (19) is the nonlinear Schrödinger equation) and solitary waves (when $N(I) = \frac{-1}{1+2I}$, the saturable nonlinearity) play a major role. They are the energy carrying modes, the "coherent structures" or "large eddies" of the Maxwell-Bloch equations. On the other hand, it is the lower amplitude phonon and radiation modes which are the source of the chaos which eventually occurs on increasing the input intensity. Details of the results are given in [26,27]. Here I sketch a few details in order to make several general remarks.

Some of the most interesting behavior occurs in the limit of large Fresnel number or small $\gamma$. One can get some idea of what to expect by imagining $A(\vec{x}) = a$ to be constant in $\vec{x}$ and solving (19) directly for $\vec{x}$ independent solutions. If $B_n(\vec{x},0) = g_n$ the map becomes the Ikeda map

$$g_{n+1} = a + Rg_n \exp i(\phi + 1/2\ N(g_n g_n^*)L_1),\tag{21}$$

whose fixed point modulus $|g|$, where $g = \lim g_n$, is graphed as function of $a$ in Figure 5. Note the bistable response of the system.

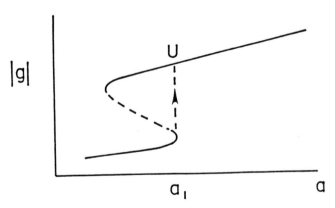

Figure 5. Locus of fixed points for plane wave map

Now, we reintroduce the dependence of the envelope on the transverse
dimension and observe that for large Fresnel numbers and moderate
gradients of the input pulse, the initial effect of diffraction is
small and therefore each position of the envelope cross-section
initially behaves as if it were a plane wave at that amplitude.

However, if the middle of the input beam $A(\vec{x})$ is greater than $a_1$

in Figure 5, the portion of the beam for which $A(\vec{x}) > a_1$ will switch

to the upper branch U whereas the outer parts of the beam will be
attracted to the lower branch. This will create sharp gradients and

even through $\gamma$ is small, the diffraction term in (19), $\gamma \nabla^2 B_n$,

becomes important. In one transverse space dimension, the balance
between diffraction and nonlinearity leads to the emergence of a
sequence of solitary waves. It turns out that the fixed point of the
infinite dimensional map is just such a sequence of solitary waves in
the beam center which sit on top of a long shelf (a small k mode)
which stretches over the full transverse width of the beam. In Figure
6a, we show a situation in which the width of the switch up region
is only wide enough for one solitary wave. The amplitude of the
solitary waves (in Figure 6a, solitary wave) and that of the shelf are
determined analytically from the map. The agreement between the
theoretical prediction and the results of numerical experiment are
very good.

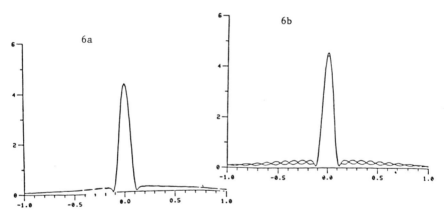

Figures 6a,b (a) The transverse profile of the fixed point of the
map (20) consisting of one solitary wave and a shelf. (b) The
situation after the shelf has period doubled. The figure is a
juxtaposition of the transverse profiles on two successive passes.


I now want to stress several important points. The first is that
the fixed point attractor consists of natural asymptotic states of the

propagation equation. This is not altogether surprising because at the end of each round trip, the fixed point signal is distorted by (20) and then has to reform itself by the time it reaches the end of the medium on the next pass. The fixed point must therefore consist of structures which propagate without change of shape, namely the asymptotic states. Otherwise the fixed point would be extremely sensitive to the length of the medium which it isn't. What is not clear is why a composition of these states should be an attractor. To be sure, the map, which, because $R < 1$, is dissipative, will and does decide what are the allowable amplitudes of the solitary waves and the shelf, the components of the (infinite dimensional) fixed point. This means that we can explain how the attractor is reached when the transverse structure only contains the asymptotic states, namely, when the initial condition belongs to a subspace spanned by these structures. But the distortion at the beginning of each pass also introduces other modes, radiation modes which somehow must disappear by the time the signal again reaches the end of the nonlinear medium. Since there is no damping in the medium, these modes must be gotten rid of in a different way. They do not damp out. Rather, they disperse away and are lost through $x = \pm \infty$ which in practice is at the distant transverse sidewalls of the cavity.

We learn, then, that when dealing with partial differential equations which support dispersive waves, we must include the dispersion of energy out of finite regions of the configuration space as an equivalent mechanism to dissipation for causing phase space volumes to collapse. Indeed, in situations where the dissipation is relatively weak, dispersion will be the principal mechanism which separates out the coherent structures (the asymptotic states) from other components of the flow field.

It is also interesting that in one space dimension, the solitary waves have an inhibiting effect in the onset of chaos. It turns out that the onset of chaos in one dimensional optically bistable cavities is a lower branch phenomenon and is triggered in the region outside the beam center and away from the solitary waves. What happens is that the lower branch shelf distabilizes through a period doubling bifurcation to a low amplitude mode with a preferred wavenumber [26]. For values of the input intensity just past the critical value, this almost sinusoidal spatial wave saturates and the new attractor is a limit cycle of period two (the structure repeats on every other pass) consisting of solitary waves in the beam center and a low amplitude sinusoidal mode on the outside. For larger amplitudes of the input intensity, the sinusoidal wave destablizes to a nearby sideband through a Benjamin-Feir like mechanism. This sideband state lies outside the range of modes to which the original shelf is unstable and therefore relaxes to a shape close to that of the original shelf. But this is unstable and the process repeats and leads to a chaotic time signal which is almost period two but which intermittently returns to the neighborhood of the unstable fixed point, the original shelf. Thus a homoclinic orbit is set up and the temporal chaos is associated with the sensitive behavior of the true orbit in its neighbourhood.

So far, therefore, the signal is temporally chaotic and thanks to dispersion, spatially coherent. We are now in the stages of addressing the question as to what happens as the stress parameters (the input intensity or the strength of the nonlinearity) are further increased. Our numerical experiments indicate that the spatial structures become more complicated (but still appear to be dominated by identifiable coherent states) and that the system passes through several windows of temporal chaos. We have no clear idea as yet what might lead to the breakdown of spatial coherence and the decay of spatial correlations. As I mentioned in the introduction this could happen either as the result of the emergence of low dimensional fractal spatial structures or the emergence of an infinite number of active modes.

A second example of where the idea of projecting the field variable into the (nonlinear) normal mode basis of a nearby integrable model is used in the forced, damped sine-Gordon equation under periodic boundary conditions:

$$u_{tt} - u_{xx} + \sin u = \Gamma \cos \omega t - \alpha u_t \, ,$$
$$u(x+L,t) = u(x,t) \, , \tag{22}$$

which is the continuum limit of a model consisting of a line of pendula hanging from a torsion wire. When all the pendula are rigidly constrained to swing together, that is when $u(x,t)$ is completely independent of $x$, the problem is equivalent to a single, damped, driven pendulum; in this case the long time behavior of $u$ as a function of the parameters $(\Gamma, \alpha, \omega)$ is well understood. In particular, chaotic states exist. However, the modulational instability makes spatial structure inevitable. Once present, these spatial structures inhibit the onset of and completely alter the route to chaos. For example, fix the initial data $u(x,t=0)$ with one localized state per period and fix the parameters $\alpha = 0.2$ and $\omega = 0.6$. Then, in order of increasing driving strength $\Gamma$, a typical bifurcation sequence is [28]: (1) the localized state decays to an x-independent state which is temporally locked to the sinusoidal river; (2) the state adjusts to one with one localized excitation per period which sits on an x-independent, flat background and which is temporally locked to the driver; (3) the state adjusts to one with two localized excitations per period which sit on an x-independent, flat background and which are temporally locked to the driver; (4) the state adjusts to one with a few localized excitations per period which sit on a flat background and which are temporally chaotic. This bifurcation sequence, in the presence of these robust spatial structures, is completely different from the bifurcation sequences for the single pendulum.

The natural question arises: can these spatially localized asymptotic states be represented accurately by a few "breathers", a temporally periodic soliton wave train of the unperturbed, integrable sine Gordon equation? Since the perturbations can be large and the

temporal behavior chaotic, the accuracy of such a representation is
uncertain a priori. One could check its accuracy indirectly by
attempting to fit a multisoliton wave train to the spatial
configurations. An alternative method is to make a direct check using
inverse spectral theory. At time t we take the profile u(x,t) and
numerically compute its spectral transform. In this spectral
representation, the soliton content of the wave is directly measured.
The results of these measurements show, that even when Γ is large
enought to produce a temporally chaotic state, the spatial structure
is accurately resolved by a low dimensional, sine-Gordon wavetrain.
For details, see [28]. Thus, in this near integrable example, the
attractors are low dimensional, consisting of robust, localized
spatial structures, which are accurately represented by the soliton
wave trains of the unperturbed equation.

IIc. <u>Almost periodic patterns</u>. The classical prototype for pattern
formation is Rayleigh-Bénard convection in a horizontal layer of fluid
heated from below. If the box has a large horizontal dimension, the
sidewalls are too distant to remove the orientational degeneracy I
spoke about in Ib. Near onset, there is a preferred wavelength but
linear stability theory does not distinguish between roll or multiroll
patterns of different orientation. Nonlinear theory, on the other
hand, can distinguish between the various planforms (whether in an
infinite horizontal geometry, the field consists of straight parallel
rolls or hexagons or rectangles). Busse and his coworkers (18) have
carried out an extensive analysis over the years on this question and
have determined the various parameter ranges in which the diverse
planforms are the preferred states. In addition, he has also
determined the domain in Rayleigh number, wavenumber, Prandtl number
space (the Busse balloon-windsock) for which straight parallel rolls
(fully nonlinear solutions of the Oberbeck-Boussinesq equations for
which $R - R_c = 0(1)$) are stable.

However, in boxes which are finite (but for which the aspect ratio
is still large), one rarely observes a convection field consisting of
straight parallel rolls. Rather one sees a mosaic of patterns which
<u>locally</u> look like patches of straight parallel rolls but whose
orientation changes by order one amounts over the box. The rolls are
spawned locally, often influenced by the horizontal boundaries which,
if in good thermal contact with the fluid, will cause the roll axis to
be locally perpendicular to the walls. Because of symmetry, each
orientation of rolls is equally allowed and the competition between
roll patches of different orientations leads to a weak turbulent
behavior even for values of the Rayleigh number close to critical. In
addition, because the container cannot be tiled by a continuous
wavevector field, defects and other discontinuities are seeded and
these structures play an important role in the rearrangement of
patterns.

In order to try to capture this behavior, one tries to average out
over the local roll structure and describe the patterns in terms of
the dynamics of a slowly varying wavevector. The starting point is

the result of Busse that there exist stable, fully nonlinear, exactly periodic solutions of the Oberbech-Boussinesq equations. These solutions are given by

$$u(\vec{x},t) = f(\theta; A,R), \quad \vec{x} = (x,y),\qquad(23)$$

where $\nabla\theta = \vec{k}$, the wavevector, is related algebraicly to the amplitude parameter $A$ through a nonlinear dispersion relation

$$\sigma(R,|\vec{k}|,A) = 0.\qquad(24)$$

The amplitude parameter $A$ is an integration constant introduced by solving for $f$ as function of $\theta$ (the partial differential equation for $u(\vec{x},t)$ becomes an ordinary differential equations for $f$ in $\theta$) and the eikonal equation (24) is found by insisting that $f$ is $2\pi$-periodic in $\theta$. The next step is to adapt this picture to the case of a finite geometry in which roll patches, with the structure (23), appear at various locations in the box. Locally, the rolls appear to be almost straight and parallel but the wavevector $\vec{k}$ changes slowly over many roll wavelengths and there can be an order one difference in wavevector from one side of the box to the other. As I have mentioned, conducting boundaries in good thermal contact with the fluid tend to induce rolls whose axis is perpendicular to the boundary. Therefore at the boundaries, the direction of $\vec{k}$ follows that of the boundary.

These kinds of solutions have been described by Cross and Newell [29] by allowing $\vec{k}$ to be a slowly varying wavevector across the box. The small parameter in the theory is not amplitude but the inverse of the aspect ratio. Using ideas developed originally by Whitham [30] in the context of slowly varying wavetrains, Cross and Newell developed an equation for the slow change of the phase gradient of the rolls. The equation has a universal character in that it depends only weakly on the original governing equations. Its predictions are consistent with many of the features observed in experiment [29,31]. One conclusion is that the rearranging patterns must take a long time to settle down. Indeed, in many situations, it would appear that the system never reaches an equilibrium state at all but lies on a low dimensional chaotic attractor. Singular solutions of the phase gradient equation, namely defects, dislocations, grain boundaries, play an important role in the rearrangement of patterns. They are topological necessities because the box cannot be tiled with a pattern corresponding to a continuous vector field $\vec{k}$. In several cases, their behavior appears to be the most active ingredient in keeping the system time dependent. At present, however, we have not been able to extract from the Cross-Newell equation a full picture of how the singular solutions (the defects, dislocations etc.) interact with the continuous field. Nevertheless, it would appear that the equation

does contain the information to analyze the long time dynamics and its
universality makes it a good candidate for continued study.

III. Large values of R. So far, we have attempted to describe
several situations in which one can obtain legitimate approximations
to the Navier-Stokes or Maxwell-Block equations which are either
finite dimensional (Ia, IIb and in rare cases IIa) or are simpler
partial differential equations of universal type (Ib, IIc). Using
these reductions, we have been able to indicate how the dynamics of
systems which may be temporally periodic or chaotic, but which are
spatially regular, are captured. What is difficult and what we have
yet failed to do is capture the onset of irregular spatial structures.
We have however suggested that the complex Ginzburg-Landau equation is
a useful model for study in this regard. It is a large aspect ratio
equation and even though it is derived for situations close to onset,
it has analogous behavior to high Reynolds number flow and in
particular has a dramatically different character in one (phase or
wimpy turbulence) and two (amplitude or macho turbulence) dimensions
similar to the constrast between two and three dimensional fluid
flows. Further in one and two dimensions, one has exact solutions for
the solitary wave and singular states (in the limit of small diffusion
$\gamma^r$ and excitation $\chi$) which appear to dominate the "long time"
dynamics. We do not have equivalent singular solutions for the three
dimensional Euler equations. (There is also an accessible
experimental situation, convection in binary mixtures, where the
equation, with its parameters lying in the interesting ranges, obtains
[32]).

Our message is that these fully nonlinear, asymptotically robust
solutions of the field equations are important. Again and again they
appear to emerge as relatively isolated coherent states not only in
numerical simulations but also in real experiments. These
asymptotically robust solutions may be solitary waves or lumplike
solutions of the field equations which exist for all time or they may
have finite time singularities. Indeed we knew of at least two
instances in which these collapsing or singular states dominate the
turbulence. The first of these is plasma turbulence near the Langmuir
frequency a situation which seems to be governed by the Zakharov
equations (a nonlinear Schrödinger-like equation for the Langmuir wave
envelope augmented with an equation for the mean). Energy is
transferred to the smaller scales by the collapsing solutions until
Landau damping becomes operative and transforms the turbulent wave
energy into fast electrons. The second example of the dominance of
singular solutions is the rearrangement of patterns in convection or
condensed matter physics, in which contexts defects, dislocations and
other singular solutions of the averaged equations appear to play
important roles.

However, in spite of the observation that long lived (their
lifetime exceeds their turnover time) eddies appear to be important in
turbulence, we have no clear idea how they fit into a general picture

and how they might be used to make contact with finite dimensional
dynamics. I should mention that the fact that they are long lived
does not necessarily contradict the notion of turbulent energy
transfer in which picture a large scale eddy of typical length size $\ell$
and velocity u will transfer a significant fraction of its energy
(in three dimensions) to a slightly smaller wavelength in its turnover
time $u^{-1}\ell$. This picture is based on linear thinking which identifies
modes or eddies with length scales just as the Fourier transform
associates an amplitude a(k) with a length $2\pi k^{-1}$. Nonlinear normal
modes can have many scales and in particular the dominant structures,
the filaments of the two dimensional nonlinear Schrodinger equation
and the rapidly stretching and mutually interacting vortices of the
three dimensional Euler equations, evolve in a self similar manner
$$u(\vec{x},t) = \lambda^{-1}(t) \ F(|\vec{x}|\lambda^{-1}(t)) \quad \text{with} \quad \lambda' > 0$$ (at least in the former
case) so that as time increases the energy is automatically associated
with smaller scales. The transfer of energy to smaller scales
presumably occurs by the repeated processes of (i) a continual
stretching of vortex filaments on the average due to mutual
self interactions and (ii) the sudden and local erosion of these
structures by instabilities of an inflexional profile nature.

As I have mentioned, the operational definition of fully developed
turbulence is that it has a broadband spatial as well as temporal
power spectrum. Most turbulence experts believe it to be a large
dimensional (many degrees of freedom) phenomenon with the energy
density (the Fourier transform of the spatial correlation function)
spread over many wavenumbers. The large scales, often called the
energy containing scales, do the transport work (mass, heat, angular
momentum). The small scales dissipate the energy and the intermediate
scales (the inertial range) contain the modes which act as catalysts
in the transfer of energy from the energy containing modes to their
viscous cemetaries. The behavior in the inertial range is believed to
be self-similar, depending only on $\varepsilon$, the viscous dissipation rate
or energy transfer rate and very little on the kinematic viscosity $\nu$.

In order to make contact with finite dimensional dynamics, one
might postulate a scenario that decomposes the field into two
components. One of these consists of a superposition of exact
nonlinear solutions to the governing equations. The parameters of
these solutions will be time dependent due to mutual interaction. The
second component will consist of a sea of smaller scale motions whose
effect on the larger scale motions is captured by spatially dependent
random coefficients in the finite set of ordinary differential
equations describing the evolution of the large eddies. A principal
effect of the small scale motions, felt through the random
coefficients, is to decorrelate points which are separated by
distances of the large eddy sizes. This idea would be a natural
extension of the notion of eddy viscosity in which one attempts to
capture the momentum transfer effects of the turbulent eddies in the
equation for the mean flow. On the other hand, the scenario I have

suggested attempts to identify more of the ordered structure in
turbulence than simply the mean field. Presumably, the character of
the statistics of the small scale eddies would be universal. There
would then be two sources of chaos. The first would arise from the
fact that the finite dimensional equations for the coherent eddies
with their coefficients replaced by mean values could have a strange
attractor. The effect of the random variation of the coefficients
would be twofold. First it would make the finite dimensional
attractor fuzzy although many of the properties of the original
skeleton such as the invariant measure and characteristic exponents
might be retained. The spatial dependence of the coefficients would
account for the decay of spatial correlations and the broadband
spatial structure.

I should emphasize that I have no concrete reasons to believe that
such a picture might be correct. Indeed, the broadband character of
the spatial correlations need not have anything to do with the onset
of a large number of new degrees of freedom. It could also be
explained on the basis of the existence of a strange attractor in a
finite set of evolution equations which determine the downstream
evolution of the spatial structures which survive. For the sake of
this argument, imagine the flow in a boundary layer in which the
downstream direction is infinite and the evolution of structures in
this coordinate is timelike. Now, suppose that the parameters which
characterize the flow, the vector $\vec{v}$ on the left hand column of table
I, not only satisfy a set of ordinary differential equations

$$\vec{v}_t = \vec{f}(\vec{v}, R), \quad \vec{v} \in R^m$$

in time $t$ but also satisfy a commuting flow

$$\vec{v}_s = \vec{g}(\vec{v}, R)$$

in some space-like variable. For example, away from the wall in a
turbulent boundary layer, the Taylor frozen flow hypothesis in which
correlations in time at a fixed position can be related to
correlations in downstream direction by considering the flow pattern
to be rigidly advected by the mean turbulent velocity, appears to be
well satisfied. Therefore if $\vec{v}$ is attracted to a strange attractor
as $t \to \infty$, it will be also asymptotic to a closely related strange
attractor in $s$. In this case, of course, $U\vec{g} = \vec{f}$ and the flows
obviously commute. The more general question as to whether one can
find a flow $\vec{g} \neq \vec{f}$ which commutes with $\vec{f}$ in the chaotic regime and
which commuting flow is also chaotic has never, to my knowledge, been
addressed. As an example, can one find a commuting flow to the
Lorenz equations in the chaotic regime? If this can happen, then
correlations of the components of $\vec{v}$ $(\langle v_i(t,s) \, v_j(t+\tau, s+\sigma)\rangle$ will

have broadband Fourier transforms in both variables. Hermann Flaschka and I are presently looking at this possibility.

Conclusion. It is fair to make the conclusion that the new ideas developed in connection with finite dimensional dynamics have something to say about the long time behavior of solutions of nonlinear partial differential equations, particularly in regimes where the spatial structure of the latter is regular. It would seem to me that the most important next question to ask is about the onset of spatial chaos. In which circumstances (if any) does the onset of a broadband spatial spectrum coincide with the onset of a broadband temporal spectrum? How do new degrees of freedom enter the picture? Does their number increase smoothly with an increase in the stress parameter or can it approach large values very quickly? When a boundary layer makes the rapid transition from the markedly smooth Tollmien-Schlicting wave instabilities to the emergence of streaks, cross-stream intermittency and bursts, has the dimension of the dynamics suddenly increased by an order of magnitude or does a more subtle evolution, in which the x (downstream) and z (cross-stream) structures become fractal, obtain? A major focus of our present work is to try to answer some of these questions and in particular understand the emergence of bursts in the onset of boundary layer turbulence. It is not unrealistic to suppose that singular solutions of the Euler equations are important because the streaks are associated with the rapid stretching of the downstream vorticity component by the shear of the mean flow.

## REFERENCES

[1] E.N. LORENZ, Deterministic nonperiodic flow, J. Atmos. Sci. 20, 130 (1967).

[2] D. RUELLE and F. TAKENS, On the nature of turbulence, Commun. Math. Phys. 20, 167-192 (1971). Note concerning this paper "On the nature of turbulence", Commun. Math. Phys. 21, 21-64 (1971).

[3] D. RUELLE, Large volume limit of the distribution of characteristic exponents in turbulence, Commun. Math. Phys. 87, 287-302 (1982). Characteristic exponents for a viscous fluid subjected to time dependent forces, Commun. Math. Phys 93, 285-300 (1984).

[4] C. FOIAS, O.P. MANLEY, R. TEMAN, and Y.M. TREVE, Asymptotic analysis of the Navier-Stokes equations, Physica 9D, 157-188 (1983).

[5] J. HYMAN and B. NICOLAENKO, Preprint (1985).

[6] S. NEWHOUSE, D. RUELLE, and F. TAKENS, Occurrence of strange axiom A Attractors near Quasiperiodic Flows on $T^m$, $m > 3$, Commun. Math. Phys. 64, 35-40 (1978).

[7] J.D. FARMER, E. **OTT**, and J.A. YORKE, The Dimension of Chaotic Attractors, Physica 7D, 153 (1983).

J.P. ECKMANN and D. RUELLE, Ergodic Theory of Chaos and Strange Attractors, Preprint 1985.

[8] A. LIBCHABER, S. **FAUVE**, and C. LAROCHE, Two Parameter Study of the Routes to Chaos, Physica 7D, 73 (1983).

[9] K. IKEDA, Multiple valued stationary state and its instability of the transmitted light by a ring cavity system, Optics, Comm. 30(2) 256-261 (1979).

H.J. CARMICHAEL, R.R. **SNAPP**, and W.C. SCHIEVE, Oscillatory instabilities leading to 'optical turbulence' in a bistable ring cavity, Phys. Rev. A 26 3408 (1982).

[10] A.P. FEIN,M.S.**HEUTMAKER,**and J.P. GOLLUB, Scaling at the Transition from Quasiperiodicity to Chaos in a Hydrodynamical System, Physica Scripta T9, 79-86 (1985).

[11] A.T. LIBCHABER, Preprint (1985).

[12] M.J. FEIGENBAUM, L.P. KADANOFF, and S.J. SHENKER, Quasiperiodicity in Dissipative Systems: A Renormalization Group Analysis, Physica 5D 370 (1982).

[13] D. RAND, S. OSTLUND, J. **SETHNA,** and E.D. SIGGIA, Universal transition from Quasiperiodicity to Chaos in Dissipative Systems, Phys. Rev. Lett. 49, 132 (1982) and Physica 8D, 303 (1983).

[14] Order in Chaos, Physica 7D.

[15] H.J. SCHUSTER, Deterministic Chaos, Physik-Verlag (1985).

[16] I have unashamedly stolen the use of the words metaphor and simile in this context from Steve Davis.

[17] J.T. STUART, On the Nonlinear Mechanics of Wave Disturbances in Stable and Unstable Parallel Flows. Part 1. The Basic Behavior in Plane Poiseuille Flow, J. Fluid Mech. 9, 353 (1960).

J. WATSON, On the Nonlinear Mechanics of Wave Disturbances in Stable and Unstable Parallel Flows. Part 2. The Development of a Solution for Plane Poiseuille Flow and for Plane Couette Flow, J. Fluid Mech. 9, 371 (1960).

[18] F. BUSSE, Nonlinear Properites of Convection, Rep. Prog. Phys. 41, 1929 (1978).

F. BUSSE, The Stability of Finite Amplitude Cellular Convection and Its Relation to an Extremal Principle, J. Fluid Mech. 30, 625 (1967).

A. SCHLUTER, D. **LORTZ,** and F. BUSSE, On the Stability of Steady Finite Amplitude Convection, J. Fluid Mech. 23, 129 (1965).

L.A. SEGEL and J.T. STUART, On the Question of the Preferred Mode in Cellular Thermal Convection, J. Fluid Mech. 13, 289 (1962).

F. BUSSE and K.E. HEIKES, Convection in a Rotating Layer; A Simple Case of Turbulence, Science 208, 173 (1980).

J.B. McLAUGHLIN and P.C. MARTIN, Transition to Turbulence in a Statically Stressed Fluid System, Phys. Rev. A 12, 186 (1975).

[19] F.H. BUSSE, Phase Turbulence in Convection Near Threshhold. Workshop on Instabilities in Continuous Media, Int. Union Geodesy and Geophysics, Venice 1984.

[20] G. AHLERS and R.P. BEHRINGER, Evolution of Turbulence from the Rayleigh Benard Instability, Phys. Rev. Lett. 40, 712 (1978).

J.P. GOLLUB and J.F. STEINMAN, Doppler Imaging of the Onset of Turbulent Convection, Phys. Rev. Lett. 47 505-508 (1981).

[21] A.C. NEWELL and J.A. WHITEHEAD, Review of the Finite Bandwidth Concept. Proc. IUTAM Symposium on "Instabilities in Continuous Systems", 284, Harrenalb (1969). Ed. H. Leipholz, Publ. Springer Berlin (1971), See also Lectures in Applied Mathematics, Vol. 15, 157, AMS (1974).

[22] A.C. NEWELL and J.A. WHITEHEAD, Finite Amplitude, Finite Bandwidth Convection, J. Fluid Mech. 38, 279 (1969).

L.A. SEGEL, Distant Sidewalls Cause Slow Amplitude Modulation of Cellular Convection, J. Fluid Mech. 38, 203 (1969).

[23] C.A. LANGE and A.C. NEWELL, A Stability Criterion for Envelope Equations, SIAM Journal of Appl. Math. 27, 441 (1974).

[24] V.E. ZAKHAROV, Collapse of Langmuir Waves, Soviet Phys. JETP 35, 908 (1972).

V.E. ZAKHAROV, R. **SAGDEEV,** and A. RUBINCHEK, Cavitons or Collapse? Preprint (1984).

[25] W.V.R. MALKUS, Nonperiodic Convection at High and Low Prandtl Number, Mem. Soc. Royale de Sci. de Liege 4, 125 (1972).

[26] D.W. McLAUGHLIN, J.V. **MOLONEY,** and A.C. NEWELL, Solitary Waves as Fixed Points of Infinite Dimensional Maps in an Optical Bistable

Ring Cavity, Phys. Rev. Lett. 51, 75 (1983). New Class of Instabilities in Passive Optical Cavities, Phys. Rev. Lett. 54, 681 (1985).

[27] J.V. MOLONEY, Coexistent attractors and new periodic cycles in a bistable ring cavity, Optics Comm. 48(6) 435-438 (1984); J.V. MOLONEY and H.M. GIBBS, Role of diffractive coupling and self-focusing in the dynamical switching of a bistable optical cavity, Phys. Rev. Lett. 48(23) 1607-1609 (1982).

[28] A.R. BISHOP, K. FESSER, P.S. LOMDAHL, and S.E. TRULLINGER, Influence of Solitons in the Initial State on Chaos in the Driven, Damped Sine-Gordon System, Physica 7D, 259 (1983).

E.A. OVERMAN II, D.W. McLAUGHLIN, and A.R. BISHOP, Coherence and Chaos in the Driven, Damped Sine-Gordon Equation: Measurement of the Soliton Spectrum, to appear Physica D (1985).

[29] M.C. CROSS and A.C. NEWELL, Convection Patterns in Large Aspect Ratio Systems, Physica 10D, 299 (1984).

[30] G.B. WHITHAM, Linear and Nonlinear Waves, Wiley Interscience (1974).

[31] D.I. MEIRON and A.C. NEWELL, The Structure of defects, Preprint 1985.

[32] H. BRAND, A.C. NEWELL, and A. OUARZEDDINI, Turbulence in Binary Mixtures, to appear 1985.

# FORMAL STABILITY OF LIQUID DROPS WITH SURFACE TENSION

D. Lewis[1], J. Marsden[1] and T. Ratiu[2]

## Abstract

A plane circular liquid drop with radius $r$, surface tension $\tau$ and rotating with angular frequency $\Omega$ is shown to be formally stable, in the sense of a positive definite second variation of a combination of conserved quantities, if $\frac{3\tau}{r^3} > (\frac{\Omega}{2})^2$. The proof is based on the Energy-Casimir method and the Hamiltonian structure of dynamic free boundary problems.

## 0. Introduction

Since the pioneering work of Arnold [1966a,b,c] on the Hamiltonian formulation of incompressible fluid dynamics and nonlinear stability of certain equilibrium planar flows, the Energy-Casimir method has been applied to a number of fluid and plasma stability problems. This method generalizes the classical $\delta W$ method primarily in its ability to deal with non-static flows; this is accomplished by the use of conserved quantities other than energy, such as angular momentum and generalized enstrophy. The reader is referred to the articles in Marsden [1984], Holm, Marsden, Ratiu and Weinstein [1985], Abarbanel, Holm, Marsden and Ratiu [1985], Holm, Marsden and Ratiu [1985], and Wan and Pulvirente [1985] for recent applications and additional references.

The general method, called the **Energy-Casimir method**, proceeds as follows: First we find a conserved quantity $C$ such that $H + C$, where $H$ is the energy, has a critical point at the equilibrium to be studied. Then the second variation $\delta^2(H + C)$ is calculated and tested for definiteness at the equilibrium. If it is definite, one refers to the equilibrium as being **formally stable**. Formal stability implies linearized stability; although many authors have claimed that it also implies nonlinear stability, it is known by example (Ball and Marsden [1984]) that additional estimates are required to justify this assertion. These are often provided by convexity estimates, as given in the aforementioned references,

[1] Department of Mathematics, University of California, Berkeley, CA 94720. Research partially supported by DOE contract DE-AT03-85ER 12097
[2] Department of Mathematics, University of Arizona, Tucson, AZ 85721. Supported by an NSF postdoctoral fellowship.

or by Sobolev type estimates, which can suffice for some semilinear equations (Marsden and Hughes [1983]) or in some one-dimensional problems such as KdV soliton stability (Benjamin [1972], Bona [1975]).

In this paper we find conditions which insure formal stability of a planar circular liquid drop of radius $r$, with surface tension $\tau$ and rotating with angular velocity $\Omega$. The surface of the drop is a free boundary. The conserved quantities used are angular momentum and generalized enstrophy. The second variation is shown to be positive definite if $\frac{3\tau}{r^3} > \left(\frac{\Omega}{2}\right)^2$. In particular, one has linearized stability in the $H^1$ norm on fluid variations and the $H^1$ norm on boundary variations under these circumstances. (In future work the questions of global existence of smooth solutions near this equilibrium solution and (rigorous) nonlinear stability will be addressed.) Formal stability for the spherical drop in three dimensions and circular shear flow in an annulus are also discussed.

The paper is organized as follows. In section one the Hamiltonian structure for our free boundary problem (see equations 1.7) is recalled. The Poisson brackets are derived by the usual procedure of reduction, as in Marsden and Weinstein [1982,1983] and Marsden, Ratiu and Weinstein [1984a,b]. These results are reviewed from Lewis, Marsden, Montgomery and Ratiu [1985]. While they are not absolutely necessary for the stability results, they provide a useful setting. In the second section the first and second variation calculations are carried out and formal stability is deduced.

The two-dimensional results presented here are closely related to those given in Sedenko and Iudovich [1978], although we obtain a less restrictive condition relating the surface tension coefficient and the angular velocity than the one given in their paper. (We cannot check their calculations since many steps are obscure or omitted; our final answers differ.) We feel, however, that our approach has the advantage of fitting into the general framework of stability analysis outlined in Holm, Marsden, Ratiu and Weinstein [1985]. Sedenko and Iudovich, following work of Arnold [1965] for fixed boundary fluids, consider relative equilibrium restricted to the "Helmholtz layer" of equivorticial flows; these layers are essentially the symplectic leaves of the Poisson manifold $\mathcal{N}$ defined below. In our argument, rather than explicitly restricting our variations to a specific layer or leaf, we introduce the generalized enstrophy functions and angular momentum as Lagrange multipliers and allow our variations to range over all tangent vectors to the space $\mathcal{N}$. Formal stability of two dimensional free boundary problems have also been considered by Artale and Salusti [1984], who consider rotational gravity waves without surface tension.

**Acknowledgements.** Conversations with Henry Abarbanel and Darryl Holm were very helpful in obtaining the results reported here.

## 1. Poisson Bracket and Equations of Motion

The dynamic variables we consider are the free boundary $\Sigma$ and the spatial velocity field $\mathbf{v}$, a divergence free vector field on the region $D_\Sigma$ bounded by $\Sigma$. The surface $\Sigma$ is an element of the set $S$ of closed curves (respectively surfaces) in $\mathbf{R}^2$ (respectively $\mathbf{R}^3$) diffeomorphic to the boundary of a reference region $D$ and enclosing the same area (respectively volume) as $D$. We let $\mathcal{N}$ denote the space of all such pairs $(\Sigma, \mathbf{v})$.

The Poisson bracket will be defined for functions $F, G : \mathcal{N} \to \mathbf{R}$ which possess **functional derivatives**, defined as follows:

i) $\frac{\delta F}{\delta \mathbf{v}}$ is a divergence free vector field on $D_\Sigma$ such that

$$D_\mathbf{v} F(\Sigma, \mathbf{v}) \cdot \delta \mathbf{v} = \int_{D_\Sigma} \langle \tfrac{\delta F}{\delta \mathbf{v}}, \delta \mathbf{v} \rangle dA, \tag{1.1}$$

where the (Fréchet) derivative $D_\mathbf{v} F$ is computed with $\Sigma$ fixed.

ii) $\frac{\delta F}{\delta \varphi}$ is the function on $\Sigma$ with zero integral given by

$$\frac{\delta F}{\delta \varphi} = \langle \tfrac{\delta F}{\delta \mathbf{v}}, \nu \rangle, \tag{1.2}$$

where $\nu$ is the unit normal to $\Sigma$. (The symbol $\varphi$ represents the potential for the gradient part of $\mathbf{v}$ in the Helmholtz, or Hodge, decomposition.)

iii) $\frac{\delta F}{\delta \Sigma}$ is a function on $\Sigma$ determined up to an additive constant as follows. A variation $\delta \Sigma$ of $\Sigma$ is identified with a function on $\Sigma$ representing the infinitesmal variation of $\Sigma$ in its normal direction. It follows from the incompressibility assumption that $\delta \Sigma$ has zero integral. The zero integral condition is dual (with respect to the $L_2$ pairing on $\Sigma$) to the additive constant ambiguity of $\frac{\delta F}{\delta \Sigma}$. We can smoothly extend $\mathbf{v}$ to a neighborhood of $\Sigma$, making it possible to fix $\mathbf{v}$ while varying $\Sigma$. Thus we can define the partial derivative $D_\Sigma F(\Sigma, \mathbf{v})$, which may be shown to be independent of the extension of $\mathbf{v}$ as long as $F$ is $C^1$ as $\mathbf{v}$ varies in the $C^1$ topology. We then let $\frac{\delta F}{\delta \Sigma}$ be the function determined up to an additive constant by

$$\int_\Sigma \frac{\delta F}{\delta \Sigma} \delta \Sigma \, ds = D_\Sigma F(\Sigma, \mathbf{v}) \cdot \delta \Sigma. \tag{1.3}$$

As an example, we compute the functional derivative with respect to $\Sigma$ of a function of the form $F(\Sigma) = \int_\Sigma f(\Sigma) \, ds$ for some smooth function $f$ of $\mathbf{x}$ defined in a neighborhood of a given $\Sigma$. Let $\Sigma_\epsilon$ be a curve in $\mathcal{S}$ with tangent vector $\delta \Sigma$ at $\Sigma$ and let $\eta_\epsilon$ be a curve in $Emb(\partial D, \mathbf{R}^2)$, the manifold of embeddings of $\partial D$ into $\mathbf{R}^2$, such that $\frac{d}{d\epsilon}|_{\epsilon=0} \eta_\epsilon = [(\delta \Sigma) \nu] \circ \eta_0$. Let $f_\epsilon : \partial D \mapsto \mathbf{R}$ be given by $f_\epsilon(X) := f(\eta_\epsilon(X))$ for $X \in D$ and let $ds_\epsilon := \eta_\epsilon^* ds$. Define $D_\Sigma F(\Sigma) \cdot \delta \Sigma := \frac{d}{d\epsilon}|_{\epsilon=0} \int_{\partial D} f_\epsilon ds_\epsilon$. The functional derivative $\frac{\delta F}{\delta \Sigma}$, if it exists, is the function modulo constants such that $\int_\Sigma \frac{\delta F}{\delta \Sigma} \cdot \delta \Sigma \, ds = D_\Sigma F(\Sigma) \cdot \delta \Sigma$. We calculate

$$
\begin{aligned}
D_\Sigma F(\Sigma) \cdot \delta \Sigma &= \frac{d}{d\epsilon}|_{\epsilon=0} \int_{\partial D} f_\epsilon(X) ds_\epsilon \\
&= \int_{\partial D} [df(\eta_0(X)) \cdot \delta \Sigma \nu(\eta_0(X)) ds_0 + (f \kappa \delta \Sigma)(\eta_0(X)) ds_0] \\
&= \int_\Sigma (\tfrac{\partial f}{\partial \nu} + f\kappa)(\mathbf{x}) \cdot \delta \Sigma(\mathbf{x}) ds
\end{aligned}
$$

which follows from the change of variables formula and the formula for the first variation of arc length. Thus, in this case,

$$\frac{\delta F}{\delta \Sigma} = \frac{\partial f}{\partial \nu} + \kappa f. \tag{1.4}$$

We now define the Poisson bracket on $\mathcal{N}$ as follows. For functions $F$ and $G$ mapping $\mathcal{N}$ to $\mathbf{R}$ and possessing functional derivatives as defined above, set

$$\{F, G\} = \int_{D_\Sigma} \left\langle \omega, \frac{\delta F}{\delta \mathbf{v}} \times \frac{\delta G}{\delta \mathbf{v}} \right\rangle dA + \int_\Sigma \left( \frac{\delta F}{\delta \Sigma} \frac{\delta G}{\delta \varphi} - \frac{\delta G}{\delta \Sigma} \frac{\delta F}{\delta \varphi} \right) ds, \tag{1.5}$$

where $\omega = \text{curl } \mathbf{v}$. For irrotational (potential) flow $\omega = 0$, and so this bracket reduces to the canonical bracket found by Zakharov [1968].

This Poisson bracket on $\mathcal{N}$ is derived from the canonical cotangent bracket on $T^*C$, where, in the two-dimensional case, $C = Emb_{vol}(D, \mathbf{R}^2)$ is the manifold of volume-preserving embeddings of a two-dimensional reference manifold $D$ into $\mathbf{R}^2$, by reduction by the group $\mathcal{G} = Diff_{vol}(D)$, the group of volume-preserving diffeomorphisms of $D$ (i.e. the group of particle relabelling transformations). Elements of $T^*C$ are pairs $(\eta, \mu)$ where $\eta : D \to \mathbf{R}^2$ is an element of $C$ and $\mu$, the momentum density, is a divergence free one form over $\eta$ ; i.e. to each reference point $X \in D$, $\mu$ assigns a one form on $\mathbf{R}^2$ based at the spatial point $x = \eta(X)$. We map $T^*C$ onto $\mathcal{N}$ by the map $\Pi_\mathcal{N} : T^*C \to \mathcal{N}$ which takes $(\eta, \mu)$ to $(\Sigma, \mathbf{v})$ such that $\Sigma = \partial(\eta(D))$ and $\langle \mathbf{v}(x), \mathbf{w}(x) \rangle = \mu(X) \cdot \mathbf{w}(x)$, for all vector fields $\mathbf{w}$ on $D_\Sigma$, where $x = \eta(X)$ and $\langle \, , \, \rangle$ is the Euclidean inner product. The map $\Pi_\mathcal{N}$ is invariant under the right action of $\mathcal{G}$ and so induces a bijection $\bar{\Pi}_\mathcal{N} : T^*C/\mathcal{G} \to \mathcal{N}$ which is a diffeomorphism in the appropriate topologies. Thus $\mathcal{N}$ inherits a Poisson structure determined by the relation

$$\{F, G\} \circ \Pi_\mathcal{N} = \{F \circ \Pi_\mathcal{N}, G \circ \Pi_\mathcal{N}\}_{T^*C}.$$

One computes the resulting bracket to be (1.5).

**Remark.** In some cases it may be necessary to use a more general Poisson bracket than that described above. While considerably more complicated, the generalized bracket has the advantage that it is defined for a larger class of functions. Of concern to us at present are the generalized enstrophy functions $C(\Sigma, \mathbf{v}) = \int_{D_\Sigma} \Phi(\omega) dA$, where $\omega$ is the vorticity, which we will use in the following stability analysis. These functions do not have functional derivatives of the form previously described.

We say that a function $F$ on $\mathcal{N}$ has **generalized functional derivatives** if there exist

i) $\frac{\delta F}{\delta \Sigma}(\Sigma, \mathbf{v})$      a function on $\Sigma$ determined up to a constant,

ii) $\frac{\hat{\delta} F}{\delta \mathbf{v}}(\Sigma, \mathbf{v})$      a divergence free vector field on $D_\Sigma$, and

iii) $\frac{\check{\delta} F}{\delta \mathbf{v}}(\Sigma, \mathbf{v})$      a vector field on $\Sigma$

such that

$$DF(\Sigma, \mathbf{v}) \cdot (\delta\Sigma, \delta\mathbf{v}) = \int_{D_\Sigma} \left\langle \frac{\hat{\delta} F}{\delta \mathbf{v}}, \delta\mathbf{v} \right\rangle dA + \int_\Sigma \left( \frac{\delta F}{\delta \Sigma} \delta\Sigma + \left\langle \frac{\check{\delta} F}{\delta \mathbf{v}}, \delta\mathbf{v} \right\rangle \right) ds$$

for all variations $(\delta\Sigma, \delta\mathbf{v})$. The functional derivatives $\frac{\hat{\delta} F}{\delta \mathbf{v}}$ and $\frac{\check{\delta} F}{\delta \mathbf{v}}$ are determined only up to the addition of a harmonic function, as may be seen by applying the divergence theorem.

The generalized bracket on $\mathcal{N}$ is

$$\{F,G\} = \int_{D_\Sigma} \left\langle \omega, \frac{\hat{\delta F}}{\delta \mathbf{v}} \times \frac{\hat{\delta G}}{\delta \mathbf{v}} \right\rangle dA$$
$$+ \int_\Sigma \Big( \left\langle \omega, \frac{\hat{\delta F}}{\delta \mathbf{v}} \times \frac{\check{\delta G}}{\delta \mathbf{v}} + \frac{\check{\delta F}}{\delta \mathbf{v}} \times \frac{\hat{\delta G}}{\delta \mathbf{v}} \right\rangle \tag{1.6}$$
$$+ \left\langle \frac{\delta F}{\delta \Sigma}\nu, \frac{\hat{\delta G}}{\delta \mathbf{v}} \right\rangle + \left\langle \nabla p_F, \frac{\hat{\delta G}}{\delta \mathbf{v}} \right\rangle - \left\langle \frac{\delta G}{\delta \Sigma}\nu, \frac{\hat{\delta F}}{\delta \mathbf{v}} \right\rangle - \left\langle \nabla p_G, \frac{\hat{\delta F}}{\delta \mathbf{v}} \right\rangle \Big) ds,$$

where $p_F$, a "pressure" associated with $\frac{\hat{\delta F}}{\delta \mathbf{v}}$, is the solution of the Dirichlet problem: $\Delta p_F = -\mathrm{div}((\nabla \mathbf{v}) \cdot \frac{\hat{\delta F}}{\delta \mathbf{v}})$, $p_F|\Sigma = \frac{\delta F}{\delta \Sigma} - \langle (\nabla \mathbf{v}) \cdot \frac{\hat{\delta F}}{\delta \mathbf{v}}, \nu \rangle$ and $(\nabla \mathbf{v}) \cdot \frac{\hat{\delta F}}{\delta \mathbf{v}}$ is determined by the relation $\langle \mathbf{u}, (\nabla \mathbf{v}) \cdot \frac{\hat{\delta F}}{\delta \mathbf{v}} \rangle = \langle (\mathbf{u} \cdot \nabla)\mathbf{v}, \frac{\hat{\delta F}}{\delta \mathbf{v}} \rangle$ for all vector fields $\mathbf{u}$ on $D_\Sigma$. Due to the non-uniqueness of the functional derivatives the generalized bracket is not well-defined for all pairs $F, G$ with functional derivatives as given above; if, however, we require that either $\frac{\delta F}{\delta \mathbf{v}}$ or $\frac{\delta G}{\delta \mathbf{v}}$ equals zero, then $\{F, G\}$ is uniquely defined. One can check that the generalized enstrophy functions have functional derivatives

$$\frac{\delta C}{\delta \Sigma} = \Phi(\omega),$$
$$\frac{\hat{\delta C}}{\delta \mathbf{v}} = \mathrm{curl}\,(\Phi'(\omega)\hat{\mathbf{z}})$$
$$\text{and} \quad \frac{\check{\delta C}}{\delta \mathbf{v}} = \Phi'(\omega)\hat{\mathbf{z}} \times \nu$$

and that they are Casimirs in the sense that $\{C, F\} = 0$ for **any** function $F$ on $\mathcal{N}$ with functional derivatives such that $\frac{\delta F}{\delta \mathbf{v}} = 0$. ( We will not use the generalized bracket for any further calculations in this paper.)

We now consider the equations of motion for a planar liquid drop consisting of an incompressible, inviscid fluid with a free boundary and forces of surface tension on the boundary and show that for the appropriate Hamiltonian $H$ and the Poisson bracket (1.5) defined above, these equations are equivalent to the relation $\dot{F} = \{F, H\}$ for all functions $F$ on $\mathcal{N}$ possessing functional derivatives. The equations of motion for an ideal fluid with a free boundary $\Sigma$ are

$$\frac{\partial \mathbf{v}}{\partial t} + (\mathbf{v} \cdot \nabla)\mathbf{v} = -\nabla p,$$
$$\frac{\partial \Sigma}{\partial t} = \langle \mathbf{v}, \nu \rangle, \tag{1.7}$$
$$\mathrm{div}\,\mathbf{v} = 0 \quad \text{and} \quad p|\Sigma = \tau \kappa,$$

where $\kappa$ is the mean curvature of the surface $\Sigma$ and $\tau$ is the surface tension coefficient, which is a numerical constant. Using notation for the two dimensional case, we take our Hamiltonian to be

$$H(\Sigma, \mathbf{v}) = \int_{D_\Sigma} \tfrac{1}{2}|\mathbf{v}|^2 dA + \tau \int_\Sigma ds. \tag{1.8}$$

The functional derivatives of $H$ are computed to be

$$\frac{\delta H}{\delta \mathbf{v}} = \mathbf{v},$$
$$\frac{\delta H}{\delta \varphi} = \left\langle \frac{\delta H}{\delta \mathbf{v}}, \nu \right\rangle = \langle \mathbf{v}, \nu \rangle, \tag{1.9}$$
$$\text{and, using (1.4),} \quad \frac{\delta H}{\delta \Sigma} = \tfrac{1}{2}|\mathbf{v}|^2 + \tau \kappa,$$

where $\frac{\delta H}{\delta \Sigma}$ is taken modulo constants. Thus for arbitrary $F$ possessing functional derivatives, (1.5) gives

$$\{F, H\} = \int_{D_\Sigma} \langle \omega, \frac{\delta F}{\delta \mathbf{v}} \times \mathbf{v} \rangle \, dA + \int_\Sigma \left( \langle \frac{\delta F}{\delta \Sigma} \mathbf{v}, \nu \rangle - \left[ \frac{1}{2} |\mathbf{v}|^2 + \tau \kappa \right] \frac{\delta F}{\delta \varphi} \right) ds$$

$$= \int_{D_\Sigma} \langle \frac{\delta F}{\delta \mathbf{v}}, \mathbf{v} \times \omega - \nabla \left( \frac{1}{2} |\mathbf{v}|^2 \right) \rangle \, dA + \int_\Sigma \left( \frac{\delta F}{\delta \Sigma} \langle \mathbf{v}, \nu \rangle - \langle \frac{\delta F}{\delta \mathbf{v}}, \tau \kappa \nu \rangle \right) ds$$

$$= \int_{D_\Sigma} \langle \frac{\delta F}{\delta \mathbf{v}}, -(\mathbf{v} \cdot \nabla)\mathbf{v} \rangle \, dA + \int_\Sigma \left( \frac{\delta F}{\delta \Sigma} \langle \mathbf{v}, \nu \rangle - \langle \frac{\delta F}{\delta \mathbf{v}}, \tau \kappa \nu \rangle \right) ds.$$

If (1.7) holds, then we find

$$\dot{F} = \int_{D_\Sigma} \langle \frac{\delta F}{\delta \mathbf{v}}, \frac{\partial \mathbf{v}}{\partial t} \rangle \, dA + \int_\Sigma \frac{\delta F}{\delta \Sigma} \frac{\partial \Sigma}{\partial t} \, ds$$

$$= \int_{D_\Sigma} \langle \frac{\delta F}{\delta \mathbf{v}}, -(\mathbf{v} \cdot \nabla)\mathbf{v} - \nabla p \rangle + \int_\Sigma \frac{\delta F}{\delta \Sigma} \langle \mathbf{v}, \nu \rangle \, ds$$

$$= \int_{D_\Sigma} \langle \frac{\delta F}{\delta \mathbf{v}}, -(\mathbf{v} \cdot \nabla)\mathbf{v} \rangle \, dA + \int_\Sigma \left( \frac{\delta F}{\delta \Sigma} \langle \mathbf{v}, \nu \rangle - \langle \frac{\delta F}{\delta \mathbf{v}}, \tau \kappa \nu \rangle \right) ds,$$

so $\dot{F} = \{F, H\}$. Conversely, $\dot{F} = \{F, H\}$ implies (1.7) by this same calculation. If $F$ has only generalized functional derivatives, $\dot{F} = \{F, H\}$ is still equivalent to (1.7), but now we use the bracket (1.8).

## 2. Stability of Two-dimensional Circular Flow

We consider the stability of the planar incompressible fluid flow such that the boundary $\Sigma_e$ is a circle of radius $r$ and the fluid is rigidly rotating with angular velocity $\Omega$. For this equilibrium solution of the equations of motion, we shall find a conserved quantity $C$ such that $H_C := H + C$ has a critical point at the equilibrium and then test for definiteness of its second variation. In infinite-dimensional systems, such as fluid flow, we have already noted that definiteness of the second variation is not sufficient to guarantee nonlinear stability, but it does imply stability under the linearized dynamics.

One class of conserved quantities consists of the Casimirs of the Poisson manifold $\mathcal{N}$, i.e. functions $C$ on $\mathcal{N}$ satisfying $\{C, F\} = 0$ for all functions $F$ for which the bracket is defined. We will make use of Casimirs of the form $C_1(\Sigma, \mathbf{v}) = \int_{D_\Sigma} \Phi(\omega) dA$, where $\Phi$ is a $C^2$ function on $\mathbf{R}^2$ and $\omega = \langle \text{curl } \mathbf{v}, \hat{\mathbf{s}} \rangle$. We will also include the angular momentum $C_2(\Sigma, \mathbf{v}) = \int_{D_\Sigma} \langle \mathbf{v} \times \mathbf{x}, \hat{\mathbf{s}} \rangle dA$. $C_2$ is the momentum map associated to the left action of the rotation group $SO(2)$ on $\mathcal{N}$. The conservation of $C_2$ is a consequence of the invariance of the Hamiltonian $H$ under the $SO(2)$ action, which implies $\dot{C}_2 = \{C_2, H\} = 0$. The inclusion of $C_2$ in the modified Hamiltonian $H_C$ allows us, roughly speaking, to view the fluid from a rotating frame with arbitrary angular velocity. Mathematically, $C_2$ enables us to cancel an otherwise troublesome cross term in the second variation of $H$. In the course of the calculation we shall also fix a translational frame.

Thus, we take our total conserved quantity to be

$$H_C(\Sigma, \mathbf{v}) = \int_{D_\Sigma} (\tfrac{1}{2}|\mathbf{v}|^2 + \mu < \mathbf{v} \times \mathbf{x}, \hat{\mathbf{s}} > + \Phi(\omega)) dA + \tau \int_\Sigma ds, \qquad (2.1)$$

where $\mu$ is a constant, as yet undetermined. Using elementary vector identities, we can rewrite (2.1) as

$$H_C(\Sigma, \mathbf{v}) = \int_{D_\Sigma} (\tfrac{1}{2}|\tilde{\mathbf{v}}|^2 - \tfrac{1}{2}\mu^2|\mathbf{x}| + \Phi(\omega)) dA + \tau \int_\Sigma ds, \qquad (2.2)$$

where $\tilde{\mathbf{v}} := \mathbf{v} - \mu\hat{\mathbf{s}} \times \mathbf{x}$. This rephrasing corresponds to viewing the fluid from a frame rotating with constant angular velocity $\mu$; $\tilde{\mathbf{v}}$ is the fluid velocity in the rotating frame.

The first variation of $H_C$ is computed to be

$$DH_C(\Sigma, \mathbf{v}) \cdot (\delta\Sigma, \delta\mathbf{v}) \qquad (2.3)$$

$$= \int_{D_\Sigma} (\langle \tilde{\mathbf{v}}, \delta\mathbf{v} \rangle + \Phi'(\omega) \cdot \langle \operatorname{curl} \delta\mathbf{v}, \hat{\mathbf{s}} \rangle) \, dA + \int_\Sigma \left( \tfrac{1}{2}|\tilde{\mathbf{v}}|^2 - \tfrac{1}{2}\mu^2|\mathbf{x}|^2 + \tau\kappa + \Phi(\omega) \right) \delta\Sigma \, ds.$$

We now consider the case where $\Sigma_e$ is a circle of radius $r$ and $\mathbf{v}_e = \frac{\Omega}{2}\hat{\mathbf{s}} \times \mathbf{x}$ for some constant $\Omega$, i.e. the equilibrium flow is rigid rotation with angular velocity $\Omega$. The circle $\Sigma_e$ has constant mean curvature $\kappa = \frac{1}{r}$. We require $DH_C$ to vanish at this equilibrium. Since $\omega_e = \langle \operatorname{curl} \mathbf{v}_e, \hat{\mathbf{s}} \rangle = \Omega$, $DH_C$ depends on $\Phi$ only through the constants $\Phi(\Omega)$ and $\Phi'(\Omega)$. If we set $\mu = \frac{\Omega}{2}$, corresponding to choosing a frame moving with the rigidly rotating fluid, then $\tilde{\mathbf{v}}_e = \mathbf{0}$, so

$$DH_C(\Sigma_e, \mathbf{v}_e) \cdot (\delta\Sigma, \delta\mathbf{v})$$

$$= \int_{D_\Sigma} \Phi'(\Omega) \cdot \langle \operatorname{curl} \delta\mathbf{v}, \hat{\mathbf{s}} \rangle \, dA + \left( -\frac{1}{2}\left(\frac{\Omega}{2}\right)^2 r^2 + \frac{\tau}{r} + \Phi(\Omega) \right) \int_\Sigma \delta\Sigma \, ds$$

$$= \int_{D_\Sigma} \Phi'(\Omega) \cdot \langle \operatorname{curl} \delta\mathbf{v}, \hat{\mathbf{s}} \rangle \, dA,$$

since $\delta\Sigma$ satisfies $\int_\Sigma \delta\Sigma \, ds = 0$. Thus $DH_C(\Sigma_e, \mathbf{v}_e) = 0$ iff $\Phi'(\Omega) = 0$. For convenience we choose $\Phi$ to be such that $\Phi(\Omega) = 0$, $\Phi'(\Omega) = 0$ and $\Phi''(\Omega) = 1$. (We choose a non-zero value for $\Phi''(\Omega)$ since it will improve our a priori estimates, as will be discussed below.)

The second variation of $H_C$ at a general point $(\Sigma, \mathbf{v})$ is calculated to be

$$D^2 H_C(\Sigma, \mathbf{v}) \cdot (\delta\Sigma, \delta\mathbf{v})^2 = \int_{D_\Sigma} \left( |\delta\mathbf{v}|^2 + \Phi''(\omega) \cdot |\operatorname{curl} \delta\mathbf{v}|^2 \right) dA$$

$$+ \int_\Sigma \Big[ 2 \left( \langle \tilde{\mathbf{v}}, \delta\mathbf{v} \rangle + \Phi'(\omega) \cdot \langle \operatorname{curl} \delta\mathbf{v}, \hat{\mathbf{s}} \rangle \right) \delta\Sigma \qquad (2.4)$$

$$+ \left( \tfrac{1}{2}|\tilde{\mathbf{v}}|^2 - \tfrac{1}{2}\mu^2|\mathbf{x}|^2 + \tau\kappa + \Phi(\omega) \right) \left( \delta^2\Sigma + \kappa\delta\Sigma^2 \right)$$

$$+ \tfrac{\partial}{\partial\nu} \left( \tfrac{1}{2}|\tilde{\mathbf{v}}|^2 - \tfrac{1}{2}\mu^2|\mathbf{x}|^2 + \Phi(\omega) \right) \delta\Sigma^2 - \tau(\Delta\delta\Sigma)\delta\Sigma - \tau\kappa^2\delta\Sigma^2 \Big] ds,$$

where $\Delta$ is the Laplacian on $\Sigma$ and $\delta^2\Sigma$ is the variation of $\delta\Sigma$ with respect to $\Sigma$ (see the earlier comments on the computation of functional derivatives with respect to $\Sigma$). The presence of the terms involving $\delta^2\Sigma$ is due to the constraints on the variations of $\Sigma$ arising from the fact that the manifold $\mathcal{S}$ of boundary curves is not a linear space; for fixed $\Sigma$ the space of $\mathbf{v}$'s on $\Sigma$ is linear, so no such $\delta^2\mathbf{v}$ term arises. The only non-obvious term in the second variation (2.4) is the derivative with respect to $\Sigma$ of the boundary term of the first variation. This derivative is computed in the following manner. Write the last term of (2.3) as follows:

$$\int_\Sigma \left(\tfrac{1}{2}|\tilde{\mathbf{v}}|^2 - \tfrac{1}{2}\mu^2|\mathbf{x}|^2 + \tau\kappa + \Phi(\omega)\right)\delta\Sigma\,ds$$
$$= \int_\Sigma \left(\tfrac{1}{2}|\tilde{\mathbf{v}}|^2 - \tfrac{1}{2}\mu^2|\mathbf{x}|^2 + \Phi(\omega)\right)\delta\Sigma\,ds + \tau\int_\Sigma \kappa\delta\Sigma\,ds.$$

Using equation (1.4) and the definition given in the general computation of $\frac{\delta F}{\delta\Sigma}$ for $F(\Sigma)$ of the form $\int_\Sigma f(\Sigma,\mathbf{x})ds$, we see that the first term of the preceeding expression has derivative with respect to $\Sigma$ given by

$$\int_\Sigma \left[\tfrac{\partial}{\partial\nu}\tfrac{1}{2}\left(|\tilde{\mathbf{v}}|^2 - \mu^2|\mathbf{x}|^2 + \Phi(\omega)\right)\delta\Sigma^2 + \tfrac{1}{2}\left(|\tilde{\mathbf{v}}|^2 - \mu^2|\mathbf{x}|^2 + \Phi(\omega)\right)\delta^2\Sigma\right.$$

$$\left.+\tfrac{1}{2}\left(|\tilde{\mathbf{v}}|^2 - \mu^2|\mathbf{x}|^2 + \Phi(\omega)\right)\kappa\delta\Sigma^2\right]ds.$$

The $\Sigma$ variation of the second integral is clearly the second variation of the arc length of $\Sigma$ with respect to $\delta\Sigma$, which may be computed to be

$$\tau\int_\Sigma [-(\Delta\delta\Sigma)\delta\Sigma + \kappa\delta^2\Sigma]ds.$$

Adding these two terms and regrouping gives expression (2.4).

For the circular flow described above the second variation reduces to

$$D^2 H_C(\Sigma_e,\mathbf{v}_e)\cdot(\delta\Sigma,\delta\mathbf{v})^2 = \int_{D_\Sigma}\left(|\delta\mathbf{v}|^2 + |\mathrm{curl}\,\delta\mathbf{v}|^2\right)dA$$

$$+\int_\Sigma\left[\left(-\tfrac{1}{2}\left(\tfrac{\Omega}{2}\right)^2 r^2 + \tfrac{\tau}{r}\right)(\delta^2\Sigma + \kappa\delta\Sigma^2) - \left(\tfrac{\Omega}{2}\right)^2 r\delta\Sigma^2 - \tau(\Delta\delta\Sigma)\delta\Sigma - \tfrac{\tau}{r^2}\delta\Sigma^2\right]ds$$

$$= \int_{D_\Sigma}\left(|\delta\mathbf{v}|^2 + |\mathrm{curl}\,\delta\mathbf{v}|^2\right)dA - \int_\Sigma\left[\left(\tfrac{\Omega}{2}\right)^2 r\delta\Sigma^2 - \tau(\Delta\delta\Sigma)\delta\Sigma - \tfrac{\tau}{r^2}\delta\Sigma^2\right]ds, \qquad (2.5)$$

since the integral $\int_\Sigma\left(\delta^2\Sigma + \kappa\delta\Sigma^2\right)ds$ is the variation with respect to $\delta\Sigma$ of $\int_\Sigma\delta\Sigma\,ds$, which is identically zero due to our restriction to area preserving variations. It follows that $D^2 H_C(\Sigma_e,\mathbf{v}_e)$ is positive-definite iff

$$\tau\int_\Sigma\left(-\tfrac{1}{r^2}\delta\Sigma^2 - (\Delta\delta\Sigma)\delta\Sigma\right)ds > \left(\tfrac{\Omega}{2}\right)^2 r\int_\Sigma\delta\Sigma^2 ds \qquad (2.6)$$

for all area preserving variations $\delta\Sigma$.

We can simplify the expression of this condition by estimating $-(\Delta\delta\Sigma)\delta\Sigma$ using eigenvalues of the negative of the Laplacian on the circle of radius $r$. The eigenfunctions are $\delta\Sigma_{k,\phi}(\theta) := \cos k(\theta - \phi)$ with eigenvalues $\lambda_{k,\phi} = \left(\frac{k}{r}\right)^2$ for all positive integers $k$ and $\phi \in [0, 2\pi)$. It is clear that the left side of (2.6) equals zero when $\delta\Sigma = \delta\Sigma_{1,\phi} = \cos(\theta - \phi)$. This eigenfunction corresponds to an infinitesmal translation in the $\phi$ direction, as $\cos(\theta - \phi)$ is the linearization of the normal perturbation $\Delta\Sigma_{\epsilon,\phi} = \epsilon\cos(\theta - \phi) + \sqrt{r^2 - \epsilon^2\sin^2(\theta - \phi)} - r$ associated to a displacement of length $\epsilon$ in the $\phi$ direction. If we wish to consider our system modulo position, regarding two configurations as equivalent if one can be obtained from the other by a Euclidean motion, then we can simply ignore the perturbations generated by the lowest eigenfunctions $\delta\Sigma_{1,\phi}$ and test for the definiteness of $D^2H_C$ only with respect to perturbations which actually distort the drop shape. In this case, taking $\lambda_{2,\phi} = \frac{4}{r^2}$ as the lowest admissible eigenvalue, it follows from (2.6) that $D^2H_C$ is positive-definite iff

$$\frac{3\tau}{r^3} > \left(\frac{\Omega}{2}\right)^2.$$

**Remarks. 1.** This procedure of ignoring Euclidean motions is equivalent to evaluating the definiteness of the second variation on the quotient space of fields $(\Sigma, \mathbf{v})$ modulo Euclidean motions; in other words, it is precisely establishing formal stability of our solution viewed as a **relative equilibrium** in the sense of Poincaré; see Marsden and Weinstein [1974] or Abraham and Marsden [1978] for the abstract theory.

**2.** The interior integral in the second variation (2.5) is equivalent to the square of the $H^1$ norm of $\delta\mathbf{v}$; had we chosen $\Phi''(\omega) = 0$ rather than $\Phi''(\omega) = 1$ this term would have equaled the square of the $L^2$ norm of $\delta\mathbf{v}$ instead. We expect that, as in the proof of global existence of two dimensional flows (Kato [1967]), this term will be useful in our investigation of nonlinear stability. A key difficulty will be to determine if the stability estimates are sufficient to prevent the breaking of small surface waves. For the somewhat related problem of vortex patches (without surface tension) it is known that surface waves can break; nevertheless one still has stability (Wan and Pulvirente [1985]).

**3.** The formal stability analysis outlined above for a circular liquid drop in $\mathbf{R}^2$ may also be applied to a spherical drop in $\mathbf{R}^3$ rotating about, for example, the $\hat{\mathbf{z}}$ axis. The generalized enstrophy functions $\Phi(\omega)$ are not conserved in the three-dimensional case and are therefore dropped from the Hamiltonian $H_C$; otherwise, the analysis procedes as in the two-dimensional case, with the following numerical differences: our curvature conventions are such that the mean curvature $\kappa$ of the sphere equals $\frac{2}{r}$, the second variation of area is given by $\tau \int_\Sigma [-(\Delta\delta\Sigma)\delta\Sigma + \frac{\kappa}{2}\delta^2\Sigma]ds$ and the first and second eigenvalues of the Laplacian on the sphere are, respectively, $\frac{2}{r^2}$ and $\frac{6}{r^2}$. Thus, in the case of the two-sphere, $D^2H_C$ is positive-definite iff

$$\frac{4\tau}{r^3} > \left(\frac{\Omega}{2}\right)^2.$$

**4.** The stability criteria found by Sedenko and Iudovich [1978] for circular shear flow in an annulus may be obtained by the methods described above; in fact, we find a less restrictive relationship between $\tau$ and $\omega_e$ than is given in their paper. We consider the equilibrium flow $\mathbf{v}_e = \frac{\omega_e(|\mathbf{x}|)}{2}\hat{\mathbf{z}} \times \mathbf{x}$ in the annulus $r_0 \leq |\mathbf{x}| \leq r_\Sigma$ with fixed inner boundary $\Sigma_0$ of radius $r_0$ and free outer boundary $\Sigma$ of radius $r_\Sigma$; $\omega_e : [r_0, r_\Sigma] \mapsto \mathbf{R}$ is a $C^1$ function with no critical points. We add the conserved quantity $\lambda \int_{\Sigma_0} \mathbf{v} \cdot dl$, where $\lambda$ is an as yet undetermined constant, to the modified Hamiltonian (2.1). Taking the first variation of $H_C$ as in (2.3) and integrating the $\Phi'(\omega)$ term by parts, we find that

$$DH_C(\Sigma, \mathbf{v}) \cdot (\delta\Sigma, \delta\mathbf{v})$$

$$= \int_{D_\Sigma} \langle \tilde{\mathbf{v}} + \mathrm{curl}(\Phi'(\omega)\hat{\mathbf{z}}), \delta\mathbf{v}\rangle \, dA - \int_\Sigma \Phi'(\omega)\delta\mathbf{v} \cdot dl - \int_{\Sigma_0} \Phi'(\omega)\delta\mathbf{v} \cdot dl + \lambda \int_{\Sigma_0} \delta\mathbf{v} \cdot dl$$

$$+ \int_\Sigma \left(\tfrac{1}{2}|\tilde{\mathbf{v}}|^2 - \tfrac{1}{2}\mu^2|\mathbf{x}|^2 + \tau\kappa + \Phi(\omega)\right)\delta\Sigma \, ds.$$

Using the techniques outlined in Holm, Marsden, Ratiu and Weinstein [1985], we find a function $\Phi$ of $\omega$ such that $\tilde{\mathbf{v}}_e = \mathrm{curl}(\Phi'(\omega_e)\hat{\mathbf{z}})$ and $\Phi'(\omega_e(r_\Sigma)) = 0$ (it is essential for this step that $\omega_e$ have no critical points). Letting $\lambda = \Phi'(\omega_e(r_0))$ and $\mu = \frac{\omega_e(r_\Sigma)}{2}$, we obtain $DH_C(\Sigma_e, \mathbf{v}_e) = 0$. The condition $\tilde{\mathbf{v}}_e = \mathrm{curl}(\Phi'(\omega_e)\hat{\mathbf{z}})$ implies

$$\Phi''(\omega_e) = \frac{|\mathbf{x}|(\omega_e(|\mathbf{x}|) - \omega_e(r_\Sigma))}{2\omega_e'(|\mathbf{x}|)}.$$

The second variation at the equilibrium point is computed to be

$$D^2 H(\Sigma_e, \mathbf{v}_e) \cdot (\delta\Sigma, \delta\mathbf{v})^2$$

$$= \int_{D_\Sigma} (|\delta\mathbf{v}|^2 + \Phi''(\omega_e) \cdot |\mathrm{curl}\,\delta\mathbf{v}|^2)dA + \int_\Sigma \left[\left(\frac{\omega_e}{2}\right)^2 |\mathbf{x}|\delta\Sigma^2 + \tau(\Delta\delta\Sigma)\delta\Sigma + \frac{\tau}{|\mathbf{x}|^2}\delta\Sigma^2\right] ds.$$

It follows that the flow is formally stable iff

$$\frac{|\mathbf{x}|(\omega_e(|\mathbf{x}|) - \omega_e(r_\Sigma))}{2\omega_e'(|\mathbf{x}|)} \geq 0 \tag{2.7}$$

$$\text{and} \qquad \frac{3\tau}{r_\Sigma^3} > \left(\frac{\omega_e(r_\Sigma)}{2}\right)^2. \tag{2.8}$$

Condition (2.7) is equivalent to the interior vorticity condition given by Sedenko and Iudovich; condition (2.8) differs from the analogous surface tension condition in Sedenko and Iudovich by a factor of three. (The derivation of this inequality from the variation of the mean curvature is not explained in their paper, so, as before, we were unable to determine the source of the difference.)

If we consider the annulus $r_\Sigma \leq |\mathbf{x}| \leq r_0$ with fixed outer ring and free inner ring, moving with velocity $\mathbf{v}_e$ as before, then the flow is formally stable iff condition (2.7) holds. The analogue of (2.8) is

$$\left(\frac{\omega(r_\Sigma)}{2}\right)^2 + \frac{3\tau}{r_\Sigma^3} > 0.$$

The case of a rigidly rotating annulus, i.e. constant $\omega_e$, is analogous to that of a rigidly rotating circle; in this case we take $\lambda = \Phi'(\omega_e) = 0$ and $\mu = \frac{\omega_e(r_\Sigma)}{2} = \frac{\omega_e}{2}$. The resulting stability condition for an annulus with fixed inner boundary and free outer boundary is (2.8), with $\omega_e(r_\Sigma)$ replaced by the constant $\omega_e$. The rigidly rotating annulus with free inner boundary and fixed outer boundary is always stable.

**References**

Abarbanel H., Holm D., Marsden J., and Ratiu T. (1985) Nonlinear stability of stratified fluid equilibria. *Phil. Trans. Roy. Soc. London* (to appear).

Arnold V.I. (1965) Variational principle for three-dimensional steady-state flows of an ideal fluid. *Prikl. Mat. and Mekh..* **29**. 846–851. (*Applied Math. and Mechanics.* **29**. 1002–1008.)

Arnold V.I. (1966a) Sur la géometrie differentielle des groupes de Lie de dimension infinie et ses applications a l'hydrodynamique des fluides parfaits. *Ann. Inst. Fourier, Grenoble.* **16**. 319–361.

Arnold V.I. (1966b) Sur un principe variationnel pour les écoulements stationaires des liquides parfaits et ses applications aux problèmes de stabilité non linéaires. *J. Mécanique.* **5**. 29–43.

Arnold V.I. (1966c) An a priori estimate in the theory of hydrodynamic stability. *Izv. Vyssh. Uchebn. Z. Math..* **54**. 3–5. (English translation: *Transl. AMS.* **19**. (1969). 267–269. See also *Dokl. Akad. Nauk.* **162**. (1965). 773–777.)

Artale V. and Salusti E. (1984) Hydrodynamic stability of rotational gravity waves. *Phys. Rev. A.* **29**. 2787–2788.

Ball J.M. and Marsden J.E. (1984) Quasiconvexity, second variations and nonlinear stability in elasticity. *Arch. Rat. Mech. An..* **86**. 251–277.

Benjamin T.B. (1972) The stability of solitary waves. *Proc. Roy. Soc. London.* **328A**. 153–183.

Bona J. (1975) On the stability theory of solitary waves. *Proc. Roy. Soc. London.* **344A**. 363–374.

Holm D., Marsden J., and Ratiu T. (1984) Nonlinear stability of the Kelvin- Stuart cat's eyes. *Proc. AMS-SIAM Summer Conference, Sante Fe (July 1984), AMS Lecture Series in Applied Mathematics.* (to appear).

Holm D., Marsden J., Ratiu T., and Weinstein A. (1985) Nonlinear stability of fluid and plasma equilibria. *Physics Reports.* **123** (1 & 2). 1–116. See also **Physics Lett. 98A.** (1983). 15–21.

Kato T. (1967) On classical solutions of the two-dimensional non-stationary Euler equation. *Arch. for Rat. Mech. and Analysis.* **25**. 188–200.

Lewis D., Marsden J., Montgomery R., and Ratiu T. The Hamiltonian structure for dynamic free boundary problems. *Physica D.* (to appear).

Marsden J.E. (ed.) (1984) **Fluids and Plasmas: Geometry and Dynamics**. Cont. Math.. AMS. Vol. **28**.

Marsden J.E. and Hughes T.J.R. (1983) **Mathematical Foundations of Elasticity**. Prentice Hall.

Marsden J.E. and Weinstein A. (1974) Reduction of symplectic manifolds with symmetry. *Rep. Math. Phys.* **5**. 121–130.

Marsden J.E. and Weinstein A. (1982) The Hamiltonian structure of the Maxwell–Vlasov equations. *Physica D.* **4**. 394–406.

Marsden J.E. and Weinstein A. (1983) Coadjoint orbits, vortices and Clebsch variables for incompressible fluids. *Physica D.* **7**. 305–323.

Marsden J.E., Ratiu T., and Weinstein A. (1984a) Semidirect products and reduction in mechanics. *Trans. Am. Math. Soc.*. **281**. 147–177.

Marsden J.E., Ratiu T., and Weinstein A. (1984b) Reduction and Hamiltonian structures on duals of semidirect product Lie algebras. *Cont. Math. AMS.* **28**. 55–100.

Sedenko V.I. and Iudovich V.I. (1978) Stability of steady flows of ideal incompressible fluid with free boundary. *Prikl. Mat. and Mekh.*. **42**. 1049. (*Applied Math. and Mechanics.* **42**. 1148–1155.)

Wan Y.H. and Pulvirente F. (1985) The nonlinear stability of circular vortex patches. *Comm. Math. Phys.*. **99**. 435–450.

Zakharov V.E. (1968) Stability of periodic waves of finite amplitude on the surface of a deep fluid. *J. Prikl. Mekh. Tekhn. Fiziki.* **9**. 86–94.

# NONLINEAR DYNAMICS AND CHAOS IN
# OSCILLATORY RAYLEIGH-BENARD CONVECTION

Robert E. Ecke, Hans Haucke, and John Wheatley

Los Alamos National Laboratory

Los Alamos, NM  87545

## ABSTRACT

Rayleigh-Bénard convection in a low Prandtl number
fluid provides an ideal model for the experimental
study of nonlinear dynamics in hydrodynamic systems.
Dilute solutions of $^3$He in superfluid $^4$He have a
low and variable Prandtl number ($0.05\rightarrow0.15$) and
their experimental properties can be measured with
high precision.  Here we present two examples.  The
first is a study of transient dynamics at a Hopf
bifurcation to a limit cycle.  Critical slowing down
is observed and characterized using a phenomenological Landau-Hopf equation.  The second is mode-
locking of two natural quasiperiodic convective
modes.  We observe an incomplete devil's staircase
and the transition to chaos.  The characteristics of
mode-locking at low nonlinearity (low Rayleigh
number) are described by a 1-D return map which
shows a tangent bifurcation to a locked state.
Close to the chaotic onset the fractal dimension of
the attractor is calculated for locked, quasiperiodic and chaotic states and the progression to
chaos can be seen to be irregular.

## 1.  Introduction

The study of problems in nonlinear dynamics encompasses an
extremely wide range of fields and systems.  Historically, hydrodynamic
systems have played a large role in the testing and demonstration of

nonlinear theories and phenomena. They are particularly interesting
from the viewpoint of being well characterizable experimentally while
presenting extreme mathematical difficulties both analytically and
numerically. Two very useful and related hydrodynamic experiments are
Taylor-Couette flow and Rayleigh-Bénard convection. In both systems
for a variety of fluids low-dimensional chaotic attractors have been
observed[1,2,3]. In convection experiments nonlinear phenomena such
as the period-doubling cascade,[4,5,6] mode locking[7,8] and
intermittent chaos[9,10] have been reported. In order to make more
detailed and quantitative comparisons to theoretical predictions it is
necessary to have systems of high stability and on which very precise
measurements can be made. We have been studying Rayleigh-Bénard
convection in dilute solutions of $^3$He in superfluid $^4$He in a
cryogenic environment which is well suited for precise measurements of
time-dependent (temporal) nonlinear instabilities. The compensating
disadvantage of inability to observe spatial modes is a serious one.
However, for low Prandtl number convection, in a small aspect ratio
geometry such as we are studying, the bifurcations are predominantly
temporal and the spatial degrees of freedom less important.

To illustrate the special features and advantages of our system,
we will discuss two interesting nonlinear problems which we have
studied. First, after steady two dimensional Rayleigh-Bénard convection becomes unstable, the flow becomes time dependent at a Hopf
bifurcation. We present data below for which the onset and transient
dynamics in the vicinity of onset are described[11,12] very well by
an analogy to mean-field phase transition formalism, treating the
magnitude of the oscillatory convective velocity as an effective order
parameter.

The second example is the phenomenon of mode-locking between two
natural convective modes. Mode-locking is a phenomenon which occurs
quite often in physical systems and has been the object of great
theoretical and experimental interest recently[13,14]. We describe
below an experimental realization of the transition from quasiperiodicity to chaos via mode-locking which in some respects can be
treated as a simple 1-D mapping analogous to the circle map. There

are however more complicated phenomena such as hysterysis and a Hopf
bifurcation of a locked state which seem to be related to a higher
dimensional character of the mapping. The general features of this
system are illustrated in figure 1 which shows the Rayleigh-Prandtl
number parameter space and the predominant convective instabilities in
that space. The first instability is a transition from diffusive
thermal conduction to stationary (time-independent) convection which
occurs at a critical Rayleigh number $R_c$. At higher Rayleigh number
a bifurcation to a limit cycle is observed. The onset of these
oscillations is strongly Prandtl number dependent as is the next
instability, a Hopf bifurcation to a quasiperiodic state. The dashed
line denoted "DISCONTINUITY" in figure 1 represents a discontinuous
change in the state of the system where the attractor, as observed in
phase space projections, suddenly changes shape. The new attractor
evolves through a sequence of mode-lockings up to the chaotic onset.
The structure of the system beyond the chaotic onset has not been
studied in detail.

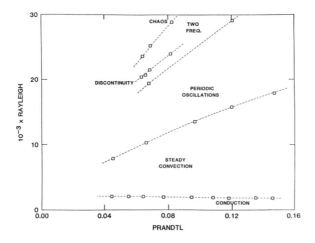

Fig. 1    Rayleigh-Prandtl number parameter space - dashed lines are
          guides to eye.

The two dimensionless parameters mentioned above which characterize the convective state are the Rayleigh number R and the Prandtl number σ. They are defined as

$$R = \frac{g\beta\Delta T d^3}{\nu\kappa} \quad \text{and} \quad \sigma = \nu/\kappa \,, \tag{1}$$

where g is the acceleration of gravity, β is the thermal expansion coefficient, d is the cell height, ν is the kinematic viscosity and κ is the thermal diffusivity. The precise definitions of these fluid parameters for $^3$He-$^4$He solutions are described elsewhere.[15]

An additional factor which determines the types of convective motion is the aspect ratio. We employ a small aspect ratio rectangular geometry with aspect ratios Γ ≡ L/2d = 1.0 and Γ′≡ W/2d = 0.70 where d = 0.80 cm is the cell height and L = 1.60 cm and W = 1.12 cm are the longer and shorter lateral sides respectively. In contrast to large aspect ratio convection where unstable roll structures are typical[16], the lateral side walls in small aspect ratio cells serve to limit the possible roll wavelengths. The roll patterns are usually two dimensional with the number of rolls dependent on the precise geometry.

In our cell there are several quite distinct stationary convective states which probably correspond to different spatial roll structures.[15,17] Here we study the state with the highest effective heat transport efficiency. Although we have no direct observations of the spatial patterns, we have some evidence that this state corresponds to two parallel rolls oriented perpendicular to the longer of the side boundaries with fluid rising in the center.

## 2. Experimental

Although an extensive description of the experimental apparatus and techniques has been presented elsewhere,[15] a brief version is given here. The RB cell consists of copper top and bottom plates, a cylindri cal stainless steel can which confines the fluid and a graphite polyi-mide-resin insert of low thermal conductance which defines the cell geometry. The geometry is rectangular with height 0.80 cm and 1.60 cm and 1.12 cm for the lengths of the longer and shorter sides.

There is a small (~4% of cell top plate area) moderately thermally insulated copper probe located at the center of the cell top plate which allows for local thermal measurements via a Au-Fe thermocouple[15,18] which is sensed by a SQUID ammeter. As fluid motion occurs in the cell, heat fluxes will be different for the probe and top plate, thereby inducing temperature differences measured by the probe thermocouple. The sensitivity of the local probe is about $10^{-7}$ K averaged over a 10 Hz bandwidth compared to a typical top-bottom temperature difference of $10^{-2}$ K.

There are germanium resistance thermometers (GRTs) on the top and bottom plates which measure absolute temperature. The bottom plate is temperature controlled, and a constant heat flow is applied to the top plate (due to the negative effective thermal expansion coefficient of the solutions[15]). The top GRT is used to determine the top-bottom temperature difference ΔT. The resolution in determining the heat flow is 5 ppm with a stability in the heat flow of order 10 ppm. Variations in bottom plate temperature regulation have limited us to an overall stability of order 50 to 100 ppm but we soon hope to improve that by as much as a factor of 10. For **time-dependent** convective states, ΔT fluctuates so that both ΔT and R are defined as time averages.

The cell was attached to a continuously operating $^3$He evaporation refrigerator whose dynamic operating range was 0.4 to 1.5 K. In practice the cell temperature was set between 0.7 and 1.1 K to obtain the desired fluid Prandtl number between 0.05 and 0.15. In the measurements presented here studying the onset of oscillations, the cell parameters are a bottom plate temperature of 0.850 K, a mean cell temperature of 0.876 K, a corresponding σ of 0.066, and a critical Rayleigh number $R_c$ for the onset of stationary convection of 2017 ± 14. In the case of the mode-locking sequence the bottom plate temperature was between 0.850 and 0.825, the mean cell temperature was 0.900 to 0.872, and the Prandtl number varied between 0.0695 and 0.0640.

## 3.  Transient Dynamics at a Hopf Bifurcation to a Limit Cycle

Primary instabilities in both Rayleigh-Bénard convection and

Taylor-Couette flow experiments can often be modeled close to onset by a nonlinear Landau-Hopf equation [11,19,20,21,22]. Here we present data for a bifurcation to oscillatory Rayleigh-Bénard convection in a dilute solution of 1.46% $^3$He in superfluid $^4$He. The steady state oscillatory amplitude dependence and the behavior of transients between steady states can be described by the same Landau-Hopf formalism. The high stability and measurement precision of our system together with some properties peculiar to the oscillatory onset allow for the determination of these properties to within $3 \times 10^{-4}$ of onset measured in reduced Rayleigh number, $(R-R_0)/R_0$, where $R_0$ is the Rayleigh number at the oscillatory onset.

The oscillatory onset occurs at $R_0/R_c = 5.129$ where $R_c = 2017$ is the critical Rayleigh number for the onset of stationary convection. The steady state rms-amplitude was previously observed by us[23] to increase as $R/R_c$ and the frequency of the oscillations to increase as $(R/R_c)^{1/2}$ with a finite value, 0.45 Hz, at onset. The spectral content of the oscillations over most of the transition region, $(R - R_0)/R_0 < 0.1$, is composed primarily of the fundamental and first harmonic.

The method we employ to measure the transient relaxation is as follows: First the oscillatory state is allowed to relax to a steady state, $R/R_c = 5.35$ above onset. Next the top plate heat flow is suddenly changed so the resultant $R/R_c$ is decreased to a value either above or below onset. We observe an initial rapid change in top-plate temperature which probably corresponds to diffusive relaxation and relaxation of the stationary roll velocity. The characteristic time $\tau_c$ for that relaxation is experimentally measured to be $2.3 \pm 0.2$ seconds and is independent of R in the region studied here. After this initial large change has decayed (typically we wait about 4 $\tau_c$) we begin recording the probe temperature amplitude response. Two such transient time series of the probe temperature oscillations are shown in figures 2(a), $R/R_c = 5.138$ (above onset) and (c) $R/R_c = 5.082$ (below onset). Previously[11] we used the oscillation envelope to characterize the transient response. An improved method, valid when the characteristic relaxation

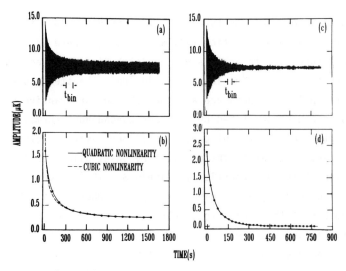

Fig. 2    Transient time series for (a) $R/R_c$ = 5.15, above onset and
(c) $R/R_c$ = 5.05, below onset.  Corresponding (b) and (d)
rms amplitudes using the 2f spectral peak and averaging times
$t_{bin}$ of 100 sec. and 30 sec. respectively.  Solid and
dashed lines are fits to the data discussed below.

time for the relaxation of the oscillations is long compared to the
period of the oscillations, is to divide the time series into small
segments and obtain via an FFT the power spectral density for each
segment.  From the power under a particular spectral peak we obtain a
narrow banded, $\delta f \sim 1$ mHz, time averaged rms-amplitude.  The
averaging time $t_{bin}$ is indicated in figure 2 for each time series.
We concentrate on the 2f harmonic because the system noise is
substantially lower at that frequency.  Although one might expect
similar behavior for the rms oscillation amplitudes obtained from the
f and 2f spectral components, this is not the case.  For now, however,
we consider only the 2f component.  The different behavior of the f
component is probably due to the way we probe the convective motion
and is described in detail elsewhere [12].

Typically, one would expect a cubic nonlinearity in a Landau-Hopf
equation describing the oscillatory velocity.  However the measurements
suggest that we measure a quantity proportional to the square of the
oscillatory velocity thus giving the observed linear dependence on R.
The appropriate dynamical equation is then

$$\frac{dA}{dt} = \alpha(R-R_0)A-\beta A^2 ,$$ (2)

where $\alpha$ and $\beta$ are positive constants, $A$ is the 2f rms oscillation amplitude and $R_0$ is $R$ at the onset of oscillations. The time integration of eqn. (2) yields

$$A(t) = \left[\gamma + \left(\frac{1}{A(0)} - \gamma\right) e^{-t/\tau}\right]^{-1} ,$$ (3)

where $\gamma \equiv \beta/\alpha(R - R_0)$ and $\tau \equiv 1/\alpha (R - R_0)$. For $R - R_0 > 0$, above onset, $\gamma = 1/A(t = \infty)$ where $A(t = \infty)$ is the steady state rms amplitude. An additional parameter, the rms system noise $N$ is necessary to fit data below onset because the measured rms amplitude decays to $N$ and not to zero. We fit to an equation of the form

$$E(t) = (A^2(t) + N^2)^{1/2} ,$$ (4)

where the noise adds as the sum of squares as expected for uncorrelated noise, $A(t)$ is the contribution described by eqn. (3) above, and $E(t)$ is the experimentally measured probe rms temperature amplitude. The data in figures 2(a) and (c) are fit by eqns. (3) and (4) and the fits are represented as solid lines in figures 2(b) and (d). An average value of $N$ of $6 \times 10^{-9}$ K is obtained from fits to all the data. The fits shown here are typical and indicate the high degree to which the dynamic equation describes the data. Fits to eqn. (4) using a cubic nonlinear piece in eqn. (2), are perceptibly inferior, see the dashed line on figure 2(b).

The steady state rms amplitude, figure 3, is obtained as a parameter in the fitting procedure. Since except very close to onset the length of the time series is always long compared to the characteristic relaxation time, this is an excellent representation of the steady state amplitude. Even close to onset the fitting seems to give good extrapolated values for the amplitude. One also obtains the characteristic relaxation time of the transient response, $\tau$, which shows a strong divergence near onset. The accuracies of $\tau$ and of the steady state amplitudes are estimated to be 5% and 1% respectively.

A divergence of the form $\tau = \tau_0 [(R-R_0)/R_0]^{-z}$ with $z=1$ is expected from eqn. (3).

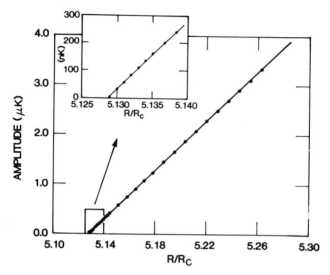

Fig. 3    Steady state rms amplitude of 2f spectral component in units of $10^{-6}$ K versus $R/R_C$.  Inset is in units of $10^{-9}$ K.

To determine $z$ and $\tau_0$ from the data we plot, figure 4, $\log_{10}$ of $\tau/\tau_\kappa$ where $\tau_\kappa \equiv d^2/\pi^2\kappa$ is the vertical thermal diffusion time versus $\log_{10}$ of $(R-R_0)/R_0$.  The data above and below onset fall on a single curve with a corresponding $z = 1.003 \pm 0.004$ and $\tau_0/\tau_\kappa = 0.247 \pm 0.003$.  The point closest to $R_0$ was not used due to its 50% uncertainty compared to 5% for the other values of $\tau$.  The value of $R_0$ which was used corresponds to the one obtained from the steady state data, $R_0 = 10344.3 \pm 5$.  $R_0$ was varied within the error bars to obtain the best agreement in terms of the data above and below onset falling on one curve.  Actually, the final value used was the steady state value given above but that is probably fortuitous considering the error bars and the fact that the data very close to $R_0$ makes the $\log_{10}$ plot very sensitive to the value of $R_0$.  Variations of as little as $\Delta R_0 \sim 0.5$ are observable.  Fitting each region separately yields values of $z^+ = 1.000 \pm 0.004$, $z^- = 1.008 \pm 0.006$,

$\tau_0^+/\tau_\kappa = 0.250 \pm 0.003$ and $\tau_0^-/\tau_\kappa = 0.242 \pm 0.005$, where the + and − refer to above and below onset respectively.

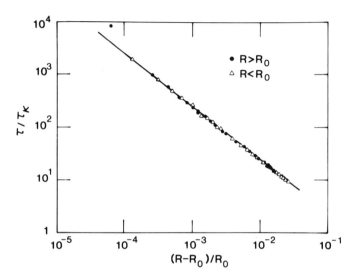

Fig. 4    Characteristic relaxation time normalized by the thermal diffusion time, $\tau/\tau_\kappa$, versus reduced Rayleigh number, $(R-R_0)/R_0$, $R_0 = 10344.3$ and $\tau_\kappa = 3.8$ s. Data points • and △ correspond to above and below onset respectively.

In summary the dynamics of the rms amplitude of the oscillatory temperature signal are well modeled by a Landau-Hopf equation with quadratic nonlinearity, eqn. (2). The exponent of the characteristic relaxation time is the same as that expected for dynamic relaxation in equilibrium critical phenomena at a mean field phase transition. This is consistent with calculations of critical fluctuations[24] near hydrodynamic instabilities which predict negligible contribution except very close to the onset.

## 4.   Quasiperiodicity, Mode-Locking and Chaos

The entrainment of nonlinear oscillators was first observed by Huyghens in the 17th century when he observed the synchronization of the frequencies of clocks coupled together by the motion of the shelf on which they sat.[25]   Mode-locking phenomena has been seen in many physical system and has been the focus of intense theoretical

interest.[13,26]. A 1-D mapping of a circle onto itself known as
the circle map has made various predictions regarding the way in which
systems mode-lock and how these lockings become "critical" at a finite
intermode coupling. At this critical coupling, the two modes are
locked at any bare winding number, defined as the ratio of the frequencies in the limit of zero coupling. These lockings form a fractal set
with fractal dimension 0.868[13,26]. Renormalization group treatments
of the circle map predict that at special irrational bare winding
numbers universal scaling relations exist.[27] Experimental
measurements[14] on Rayleigh-Benard convection systems show this
predicted scaling of the power spectral density at one of these
special irrationals, the "golden mean." Other properties are also
observed which match predictions of the circle map. In these
Rayleigh-Bénard experiments[14] one of the frequencies corresponds
to a natural convective mode while the other is applied externally.
This allows precise control of the bare winding number and of the
coupling parameter. However, the natural coupling of two intrinsic
oscillatory hydrodynamic modes has not been studied in detail. Here
we present aspects of an experimental study of two natural oscillatory
convective modes in a Rayleigh-Bénard cell containing a dilute solution
of 1.46% $^3$He in superfluid $^4$He. A more complete description will
be presented elsewhere.[8]

The Rayleigh-Prandtl number space, figure 1, shows the evolution
of the convective motion as R is increased. The limit cycle state
bifurcates to a quasiperiodic (two incommensurate frequency) state
followed by a discontinuous change in the convective motion; the phase
portrait of the attractor changes shape dramatically. The new
attractor remains quasiperiodic and continuous over a sizable range of
R before becoming chaotic. Between the onset of the new attractor and
the chaotic onset the two frequency modes undergo an entire sequence
of mode lockings at various rational integer ratios. These mode-
lockings are presented in figure 5 where the ratio of the two
fundamental frequencies, the winding number W, is plotted versus
$R/R_c$ where $R_c$ = 2017. The value of $\sigma$ for this data set is about
0.069. The figure is representative of an incomplete devil's

staircase with the additional feature that as R increases so too does
the nonlinearity of the system. This means that at larger relative R

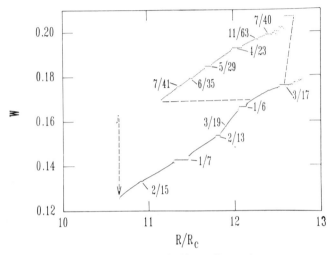

Fig. 5    The winding number, $W \equiv f_2/f_1$, plotted versus
normalized Rayleigh number.  $T_{btm} = 0.850$ K and $\sigma = 0.069$.

values there is a higher portion of locked to unlocked regions.  At low
relative R or nonlinearity a 1-D circle map gives a good description
of the state of the system.  However, for states closer to the chaotic
onset,  **hysteresis**  and Hopf bifurcations of the locked states can occur.
The Poincaré sections of these states can show stretching and folding
of the torus.  These effects imply that a 2-D mapping is necessary to
describe the states at higher relative nonlinearity.[8]
        Before discussing the 2-D mapping effects, we present the simpler
1-D map description.  For $R/R_c = 10.568$ and $\sigma = 0.067$, the phase
portrait of the attractor appears as in figure 6.  By taking a cut, as
indicated in the figure, perpendicular to the flow we obtain the
Poincaré section in figure 7.  Parameterizing this section by an angle
variable we obtain a 1-D return map, figure 8, similar to a circle
map.  One difference is that we have a $4\pi$ rotation invariance as
opposed to the usual $2\pi$ value.  In the circle map, close to a locked
state where the winding number is p/q, the qth iterate of the map will

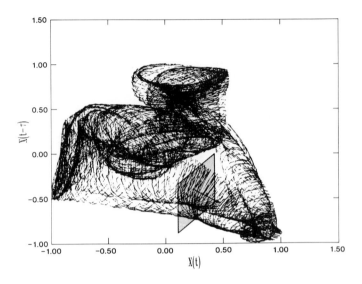

Fig. 6   Phase projection of attractor with indication of cutting plane for Poincaré sections.   τ = 120.

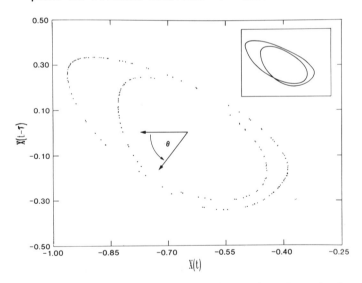

Fig. 7   Poincaré section of system trajectories through plane shown in fig. 6.   In this quasiperiodic state the phase angle wraps around to show the section of a double looped two torus. Inset illustrates how torus would fill in given more points.

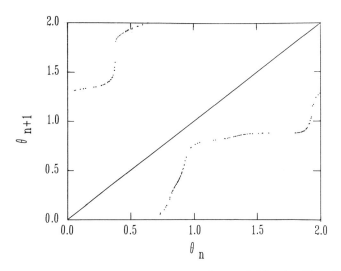

Fig. 8    Return map generated from angle variable in fig. 7 of the form $\theta_{n+1} = f(\theta_n)$.

approach a tangent bifurcation.[28]  We can demonstrate similar behavior for our data.

Figure 9 shows the 3rd iterate of the map for three states with $R/R_c = 10.568$, $R/R_c = 10.581$ and $R/R_c = 10.594$ where these correspond to a locked state and quasiperiodic states respectively.  The locked state has a winding number of 3/23 as defined experimentally W = $f_2/f_1$.  In the locked section there are three dots indicating that we need to think of q = 3 here.  The return map for the first quasiperiodic state appears to be approaching tangent bifurcations at all three of the locked points.  The other apparent bifurcation point with no corresponding locked point seems to be stagnant.  In other words successive experimental quasiperiodic states further from the locked state show that regions in the 3rd iterate which correspond to the tangencies for the locked points move dramatically while the other does not change much, see upper part of figure 9.  The fact that we experimentally define W = 3/23 yet obtain a return map consistent with q = 3 where W = p/q is we believe due in part to ambiguity in interpretation of the circle map with respect to actual flows in phase space. We discuss this point in detail elsewhere[8].

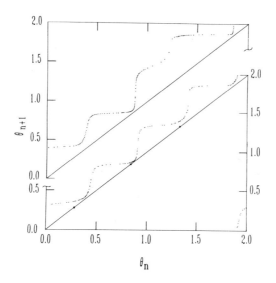

Fig. 9    Return maps for the 3rd iterate of two quasiperiodic states
(figs. 7 and 8) and a locked state which are close in
Rayleigh number. The winding number for the locked state is
3/23. This sequence shows how the mode-locking arises from
tangent bifurcations of the return map.

Other aspects of the circle map seem to apply in a quite general
way to the experimental data over the full range of $R/R_c$. The Farey
tree ordering associated with the circle map[25] can be seen in the
data, figure 5. Between any two lockings with ratios p/q and p'/q'
the predominant locking occurs at the Farey sum (p + p')/(q + q'), of
the two outer lockings.

Another general concept from the circle map is that there is some
nonlinear coupling constant, K, which controls the degree of mode-
locking between the different frequency modes. As K increases the
mode-locked regions expand in size until at K = 1 all states are mode-
locked. The resulting structure is a "devil's staircase" with a
distinct fractal structure[13]. The increase of the mode-locked
intervals with K known as Arnold Tongues is observed in our experiment.[8] We vary both the Rayleigh and Prandtl numbers and obtain
larger mode-locked regions (measured in units of $R/R_c$) as the Prandtl

number decreases.

For relatively large values of $R/R_c$ near the chaotic onset the dynamics of the system becomes increasingly more complex; a 1-D mapping can no longer describe the dynamics. We observe hysterysis near mode-locked regions depending on whether one increases or decreases R. The **hysteresis** is substantial, is not due to relaxation effects, and involves coexisting locked and quasiperiodic states. In the same region, a locked state with winding number $W = f_2/f_1 = 2/13$ undergoes a Hopf bifurcation to a quasiperiodic state with a new frequency $f'$. Figure 10 shows a sequence of Poincaré sections which illustrate this exchange of stability between $f_2$ and $f'$. The first state, figure 10(a) at $R/R_c = 11.733$, is quasiperiodic with apparent stretching and wrinkling of the two torus. The system locks with $f_2/f_1 = 2/13$ but simultaneously bifurcates to a "sublocked" state at $R/R_c = 11.748$ with $f'/f_2 = 1/5$ which yields the 10 fixed points in figure 10(b). As $R/R_c$ increases further the $f'$ mode unlocks and yields a new quasiperiodic state, figure 10(c). Finally, figure 10(d), the amplitude of $f'$ which has been steadily decreasing becomes very

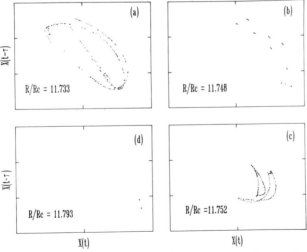

Fig. 10   Sequence of Poincaré section at successive values of $R/R_c$ indicating (a) quasiperiodic state with wrinkling, (b) Hopf bifurcated state with $f'/f_2 = 1/5$ and $f_2/f_1 = 2/13$, (c) Hopf bifurcated quasiperiodic state and (d) locked state $f_2/f_1 = 2/13$.

small and the section reflects only the $f_2/f_1$ = 2/13 locking, $R/R_c$ = 11.793. This phenomena has been experimentally observed by others[29] and reflects the 2-D properties of the return map [8,30].

At the end of the stable mode-locking region for $\sigma$ = 0.065, $R/R_c$ ~ 12, chaotic states appear. Associated with the transition to chaos is a period doubling sequence of a locked state for which two doublings are resolved. We have quantitatively characterized the transition to chaos by calculating the fractal dimension (correlation dimension) of the phase space attractor using the algorithm of Grassberger and Procaccia[31]. The data sets consist of 33 x 10^4 points for which 500 points are chosen at random in the calculation of the dimension. The sampling time is 70 msec and the time delay used to construct the phase space is 1.05 seconds. The algorithm distinguished locked, unlocked and chaotic states quite well. Figure 11 shows the dimension as $R/R_c$ is increased; locked states give values close to 1, quasiperiodic close to 2 with larger uncertainty and chaotic states between 2.5 and 3.0. The structure of the dimension shows that there is fine scale structure which reflects similar structure in the Lyapunov spectra. Examples of sensitive dependence on parameters are seen in simple dynamical systems[32].

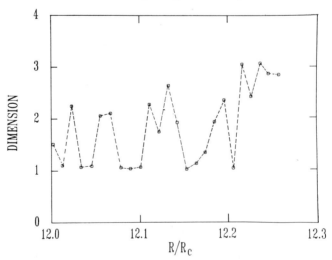

Fig. 11  Correlation dimension of sequence of states near the chaotic onset. The irregular structure is due to fine scale locking unlocking and finally chaos as $R/R_c$ increases.

5. Acknowledgments

We gratefully acknowledge discussions with J. D. Farmer, K. Kaneko and especially G. Mayer-Kress regarding the mode-locking phenomena. We also thank Y. Maeno for help on the transient dynamics. Support for this work was provided by the U. S. Department of Energy.

6. References

1. A. Brändstater, J. Swift, H. Swinney, A. Wolf, J. D. Farmer, E. Jen, and P. J. Crutchfield, Phys. Rev. Lett. 51, 1442 (1983).

2. B. Malraison, P. Atten, P. Berge, and M. Dubois, J. Phys. Lett. 44, 897 (1983).

3. M. Giglio, S. Musazzi, and V. Perini, Phys. Rev. Lett. 53, 2402 (1984).

4. A. Libchaber and J. Maurer, J. de Phys. 41, C3-51 (1980).

5. J. P. Gollub, S. V. Benson, and J. F. Steinman, Annals of New York Academy of Sciences, 357, 22 (1980).

6. H. Haucke, Y. Maeno, and J. C. Wheatley, Proceedings of LT-17, ed. by U. Eckern, A. Schmid, W. Weber and H. Wühl, (North Holland, Amsterdam, 1984), p. 1123.

7. J. P. Gollub, E. J. Romer, and J. E. Socolar, J. Stat. Phys. 23, 321 (1980).

8. H. Haucke, R. E. Ecke, and J. C. Wheatley, to be published.

9. H. Haucke, R. E. Ecke, Y. Maeno, and J. C. Wheatley, Phys. Rev. Lett. 53, 2090 (1984).

10. J. Maurer and A. Libchaber, J. de Phys. Lett. 41, L-515 (1980).

11. R. E. Ecke, Y. Maeno, H. Haucke, and J. C. Wheatley, Phys. Rev. Lett. 53, 1567 (1984).

12. R. E. Ecke, H. Haucke, Y. Maeno, and J. C. Wheatley, to be published.

13. M. H. Jensen, P. Bak, and T. Bohr, Phys. Rev. Lett. 50, 1637 (1983); D. G. Aronson, M. A. Chory, G. R. Hull, and R. P. McGehee, Comm. Math Phys. 83, 303 (1982).

14. A. P. Fein, M. S. Heutmaker, and J. P. Gollub, Physica Scripta, T9, 79 (1984); J. Stavans, F. Heslot and A. Libchaber, to be published and A. Libchaber, these proceedings.

15. Y. Maeno, H. Haucke, R. E. Ecke, and J. C. Wheatley, J. Low Temp. Phys. 59, 305 (1985).

16. M. S. Heutmaker, P. N. Fraenkel, and J. P. Gollub, Phys. Rev. Lett. 54, 1369 (1985); G. Ahlers, D. S. Cannell and V. Steinberg, Phys. Rev. Lett. 54, 1373 (1985).

17. Y. Maeno, H. Haucke, R. Ecke, and J. C. Wheatley, Proceedings of LT-17, ed. by U. Eckern, A. Schmid, W. Weber, and H. Wühl, (North Holland, Amsterdam, 1984), p. 1125.

18. Y. Maeno, H. Haucke, and J. C. Wheatley, Rev. Sci. Inst. 54, 946 (1983).

19. R. Behringer and G. Ahlers, Phys. Lett. 62A, 329 (1977).

20. J. Weisfreid, Y. Pomeau, M. Dubois, C. Normand, and P. Berge, J. Phys. Lett. (PARIS) 39, L725 (1978).

21. J. P. Gollub and M. H. Freilich, in "Fluctuations, Instabilities and Phase Transitions", ed. by R. Riste (Plenum, NY, 1975) p. 195.

22. G. Pfister and U. Gerdts, Phys. Lett. 83A, 23 (1981).

23. Y. Maeno, H. Haucke and J. Wheatley, Phys. Rev. Lett. 54, 340 (1985).

24. R. Graham, Phys. Rev. A10, 1762 (1974).

25. A. B. Pippard, The Physics of Vibration, Vol. 1, (Cambridge University Press, Cambridge, 1978).

26. P. Cvitanović, M. H. Jensen, L. P. Kadanoff, and I. Procaccia, Phys. Rev. Lett. 55, 343 (1985); P. Cvitanović, reprint.

27. M. J. Feigenbaum, L. P. Kadanoff, and S. J. Shenker, Physica 5D, 370 (1982); D. Rand, S. Ostlund, J. Sethna, and E. Siggia, Phys. Rev. Lett. 49, 132 (1982) and Physica 6D, 303 (1984).

28. K. Kaneko, Prog. Theor. Phys. 68, 669 (1982).

29. R. Van Buskirk and Carson Jeffries, Phys. Rev. A31, 3332 (1985).

30. J. Froyland, Physica 8D, 423 (1983).

31. P. Grassberger and I. Procaccia, Physica 13D, 34 (1984).

32. J. D. Farmer, Phys. Rev. Lett. 55, 351 (1985).

# NONLINEAR DYNAMICS IN SEMICONDUCTORS, INTERMITTENCY, AND LOW FREQUENCY NOISE

R. M. Westervelt, S. W. Teitsworth, and E. G. Gwinn

Division of Applied Sciences and Department of Physics
Harvard University, Cambridge, MA 02138

## ABSTRACT

An experimental and theoretical analysis is presented of extrinsic Ge photoconductors as driven nonlinear oscillators. Subharmonic and chaotic response characteristic of nonlinear dynamical systems is found in both experiments and simulations. Intermittent phenomena arising from crises and noise-induced transitions between multiple attractors are also investigated in simulations of the damped driven pendulum. Fractal basin boundaries found for the driven pendulum are shown to have important consequences for both intrinsic and noise induced intermittency. Unresolved issues and suggested experiments are given for both cases.

## 1. Introduction

Deterministic nonlinear oscillators are known to produce complex and apparantly random responses in a wide variety of theoretical and experimental systems.[1,2] However, relatively little is known about the complex nonlinear response of semiconductors, despite their practical importance. Recently, characteristic subharmonic and chaotic response has been experimentally identified in

extrinsic photoconductors[3-6] made from Ge, GaAs, and InSb.
This deterministic noise source limits the sensitivity of
these devices, which are the best far-infrared
photodetectors currently availible.[7] In section 2 we
present experimental[8,9] and theoretical[10,11] results which
show in detail how deterministic noise is a direct
consequence of the nonlinear nature of carrier generation
and trapping in extrinsic photoconductors.

Another basic property of dissipative nonlinear systems
is that they can possess multiple steady states, or
attractors, for the same parameter values. The study of how
separate attractors gain and lose stability has been the
subject of much recent work, by Pomeau and Manneville[12,13],
by Grebogi, Ott, and Yorke[14-17], and by others[18-21].
Hopping to and from nearly stable attractors produces
intrinsic intermittency, qualitative changes in the motion
on long time scales, and low frequency noise. In chaotic
systems the boundaries separating different basins of
attraction can be fractal[22,23]; fractal basin boundaries
tend to favor noise-induced hopping between multiple
attractors, and can result in extraordinary sensitivity when
the fractal dimension is large[17]. In section 3 we present
numerical simulations of the damped driven pendulum which
show in detail how intermittency and low frequency noise are
produced in this system by multiple attractors and fractal
basin boundaries, for both the intrinsic and noise-induced
cases.

## 2. Deterministic Noise in Extrinsic Photoconductors

In this section we describe experiments and simulations
of high quality extrinsic far-infrared photoconductors[7]
fabricated by E.E. Haller of Lawrence Berkeley Laboratory
from p-type Ge. Over 20 devices were studied: for all only
a single shallow acceptor level with binding energy ~10 meV

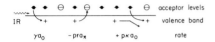

conduction band

donor levels (empty)

acceptor levels

valence band

rate

IR

$\gamma a_0$    $-pra_\star$    $+p\kappa a_0$

Fig. 1. Schematic energy band diagram of an extrinsic photoconductor illustrating the processes of carrier generation, recombination, and impact ionization (from Ref. 11).

was active, and all had B ion-implanted contacts which act as ideal resevoir contacts for holes at low temperatures. The samples were cooled to liquid He temperatures inside a cold radiation shield and uniformly illuminated by a black-body radiator with adjustable temperature T ~ 30K. When the complex nonlinear response described below was not present, the performance of these devices as photodetectors was excellent. We emphasize that the deterministic noise discussed is a fundamental property of extrinsic photoconductors which is present even for ideal devices.

Figure 1 illustrates the three processes which determine the hole concentration in extrinsic photoconductors; carrier generation, recombination, and impact ionization[7,11]. Neglecting possible space charge effects, the hole concentration p is determined by the sum of these rates, and the electric field E by Ampere's Law[11]:

$$dp/dt = \gamma a_0 - p(ra_\star - \kappa a_0),\tag{1}$$

$$dE/dt = (1/\varepsilon)(J_{ext} - pev).\tag{2}$$

Here a, $a_0$, and $a_\star = a - a_0$ are the total, neutral, and ionized acceptor concentrations; $\gamma$, r, and $\kappa$ determine the rates of generation, recombination and impact ionization; $\varepsilon$, e, and v are the dielectric constant, electronic charge, and drift velocity; and $J_{ext}$ is the measured current density.

Ordinarily the hole concentration p determined by Eq. (1) relaxes exponentially to the steady-state solution $p_s = \gamma \tau_0$,

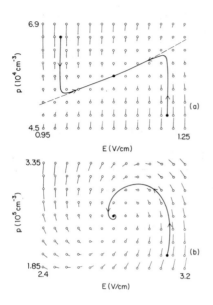

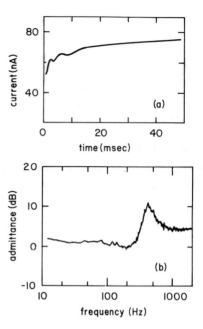

Fig. 2. Phase space flow computed from Eqs. 1 and 2 for two increasing values of current with corresponding quality factors: (a) $J_{ext}=10nA/cm^2$, $Q=0.50$; (b) $J_{ext}=100$ $nA/cm^2$, $Q=1.95$ (from Ref. 11).

Fig. 3. (a) Current transient response following a step in voltage bias to $V_{dc}=0.48V$ at T=4.2K for moderate FIR illumination (blackbody temperature 28K). (b) Small-signal admittance for the same device at T=4.2 K for $V_{dc}=0.67$ V, $I_{dc}=140nA$.

where $\tau_o = (ra_* - \kappa a_o)^{-1}$ is the lifetime against capture by ionized acceptors. However, carrier heating occurs in cooled Ge detectors at modest electric fields E > 0.1 V/cm; for larger fields the coefficients $\kappa$ and r are strongly nonlinear functions of E. This dependence couples Eqs. (1) and (2), and produces a damped nonlinear oscillator[11]. Figure 2 illustrates the transition in the transient response to a step in illumination from an overdamped exponential decay, with overshoot, to an underdamped ringing, as the d.c. bias current increases. The variables

p and E constitute the phase space for this dynamical
system, and the transients shown illustrate the approach to
the point attractors as shown in Fig. 2. The resonant
frequency $\omega_o$ and quality factor Q for the linearized case
are determined by the lifetime $\tau_o$, the dielectric relaxation
time $\tau_\rho$, and the derivatives dr/dE and d$\kappa$/dE at the bias
point, and are given in Ref. 11.

Qualitatively similar transient response anomalies,
including overshoot and damped ringing, are commonly observed
in cooled photoconductors. Figure 3a illustrates the
experimental transient response of a cooled Ge sample to a
step in voltage bias for constant infrared illumination;
damped ringing is clearly present. Figure 3b shows a
corresponding resonant peak near 500 Hz, the linearized
frequency response of the same device measured using a
network analyzer. The experimentally observed time scales
for these anomalies tend to be factors 10 to 1000 longer
than predicted by the simple spatially uniform model above.
This can be understood as a consequence of the formation of
a space charge layer trapped near the positive contact,
analogous to a virtual cathode in vacuum tubes. This
trapped charge determines the steady-state electric field in
the bulk of the sample via Gauss's Law, but can respond only
very slowly to changes in illumination or bias[11].

Driven nonlinear oscillators commonly show complex
subharmonic and chaotic response even when the undriven
transient response is simple; a classic example is the
driven pendulum. In use, photoconductors are usually
periodically driven by a chopped illumination signal which
produces a corresponding current. To theoretically test[10]
the response of the simple rate equation model above in
simulations, we added a sinusoidal component to the current:

$$J_{ext} = J_{dc} + \Delta J \cos \omega_d t, \qquad\qquad (3)$$

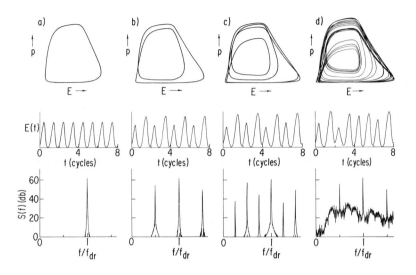

Fig. 4. Theoretical phase portrait, time record, and power spectrum for drive frequency $\omega_d/\omega_o = 1.45$ and four increasing drive amplitudes: (a) $\Delta J/J_{dc} = 1.10$; (b) $\Delta J/J_{dc} = 1.25$; (c) $\Delta J/J_{dc} = 1.35$; (d) $\Delta J/J_{dc} = 1.45$. For power spectra 0 dB is $1 \times 10^{-7} \, V^2 cm^{-4} Hz^{-1}$ (from Ref. 10).

where $J_{dc}$ is a constant. Examples of the resulting response[10] are shown in Fig. 4 where the phase diagram, time record, and power spectrum are plotted for four increasing values of the drive amplitude $\Delta J$; the drive frequency was chosen to be $\omega_d = 1.45 \, \omega_o$. Figure 4 clearly illustrates a period-doubling cascade to chaotic oscillation in all three types of plot. Poincaré sections for these time series show[10] the folding characteristic of chaotic systems, and the return map $E_{n+1}$ vs. $E_n$ on successive cycles exhibits[10] a broad maximum analogous to the logistic map. The effective noise temperature $T_{eff}$ of the chaotic power spectrum in Fig. 4 is $T_{eff} \sim 10^8$ K to $10^9$ K, many orders of magnitude larger than the noise temperatures achieved in actual devices. Thus, even a very small chaotic contribution can destroy the performance of a photoconductor.

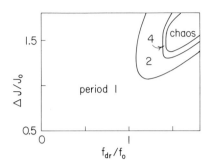

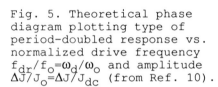

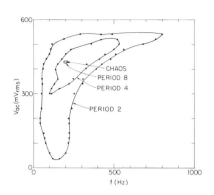

Fig. 5. Theoretical phase diagram plotting type of period-doubled response vs. normalized drive frequency $f_{dr}/f_o = \omega_d/\omega_o$ and amplitude $\Delta J/J_o = \Delta J/J_{dc}$ (from Ref. 10).

Fig. 6. Experimental phase diagram for $V_{dc}$=0.9V at T=4.2K (from Ref. 8).

Figure 5 is the "phase diagram" for a driven extrinsic photoconductor in which the type of response, period 1, 2, 4, or chaos, is plotted vs. drive amplitude and frequency[10]. As shown, the chaotic region forms a tongue near the resonant frequency which is surrounded by rings of period-doubled response; for other drive frquencies and amplitudes the response is simple and noise-free. Thus, the poorest noise performance can occur in the middle of the operating region, close to parameter values which yield excellent results. For very large drive amplitudes (not shown) the response undergoes inverse period doublings and returns to period 1. As shown in Fig. 5, the period-doubling sequence observed depends on the path taken in parameter space, but the qualitative features are in overall agreement with the logistic map, as expected from the folding found in the Poincaré sections and return map.

A series of experiments[8,9] which correspond to the simulations[10] above were conducted by driving a noise-free Ge photoconductor with a sinusoidal applied voltage of amplitude $V_{ac}$ in addition to the d.c. bias voltage $V_{dc}$. The infrared illumination was held constant, and the illumination level and d.c. voltage were chosen to give damped ringing in the transient response, as shown in Fig. 3. An example of the experimentally measured "phase diagram" is shown in Fig. 6. As shown, a chaotic island is surrounded by successive rings of period-doubled response at period 8, 4, and 2; for very large drive amplitudes, the response becomes simple again at period 1. These features are qualitatively identical to theory as in Fig. 5. The absolute values of the relative drive amplitude $V_{ac}/V_{dc}$ and frequency $2\pi f/\omega_o$ are also in reasonable agreement with Fig. 5. Because the quantitative values of the theoretical coefficients r and $\kappa$ are not accurately known, and the number of potentially adjustible parameters is large, we chose to make a comparison with experiment without adjustable parameters, using the best theoretical and experimental estimates available from the literature. Thus, this level of agreement is quite good.

A series of experimental power spectra[8,9] for increasing a.c. drive amplitudes entering the chaotic island in the phase diagram is shown in Fig. 7; note the excellent dynamic range > 80 dB, limited by our spectrum analyzer rather than by the Ge device. As shown, the response evolves from period 4, to period 8, to a chaotic "period 4" corresponding to a 4 banded chaotic attractor. The effective noise temperature of the chaotic response in Fig. 4c is ~ $10^7$ K, comparable to simulations and large enough to destroy the performance of this device as a photodetector.

In the paragraphs above we have shown, both in theory[10,11] and experiment[3,8,9], that extrinsic photoconductors form nonlinear oscillators. When sinusoidally driven, these semiconductor oscillators can produce complex subharmonic and chaotic response even though

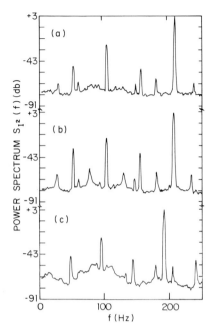

Fig. 7. Experimental current power spectra illustrating the transition to chaos in Fig. 6 as the drive frequency $f_{dr}$ is increased for $V_{dc}=0.9V$, $V_{ac}= 0.4V$:
(a) $f_{dr}$ = 209 Hz, period 4;
(b) $f_{dr}$ = 206 Hz, period 8;
(c) $f_{dr}$ = 193 Hz, 4 banded chaos (from Ref. 8)

their transient response shows only damped ringing. Many other types of semiconductor oscillators exist[24]; in many cases they form the basis of operation for a practical device[25]. Examples of oscillatory mechanisms (and devices) are: 1)transport in high electric fields - impurity impact ionization (extrinsic photoconductors), band to band impact ionization (IMPATT microwave oscillators), Zener breakdown, transferred electron effects and negative differential mobility (Gunn diode microwave oscillators); 2) nonlinear electronic networks - analog phenomena in digital computers, content addressable memory; 3)plasma oscillations - ultra-high frequency devices, magnetoplasma oscillations; 4)charge-density-wave transport. Although the semiconductor physics of these devices is often well understood as a result of work done in the 1950's and 1960's, the nonlinear dynamics and subharmonic response of these semiconductor oscillators remains practically unexplored from a modern point of view. Considerable progress could be made by the straightforward application of the modern mathematical understanding of nonlinear oscillators.

3. Intermittency and Low-Frequency Noise in the Damped Driven Pendulum

Nonlinear dynamical systems commonly possess multiple attractors for a given set of parameter values; when external noise or an intrinsic transition such as a crisis[14-16] causes individual attractors to lose stability, intermittent qualitative changes in motion occur on long time scales. We chose to study[26-28] intermittency due to multiple attractors in the damped driven pendulum, because this is a classic nonlinear system which models many physical phenomena including radio-frequency driven Josephson junctions[29-40].

The equation of motion for the damped driven pendulum written in reduced units[32] is:

$$d^2\theta/dt^2 + (1/Q)d\theta/dt + \sin(\theta) = g\cos(\omega_o t) + \delta g(t). \quad (4)$$

Here $\theta$ is the pendulum angle, $Q$ is the quality factor, $g$ and $\omega_o$ are the normalized driving torque amplitude and frequency, and $\delta g(t)$ is an external noise torque which is taken to be zero unless stated otherwise. Equation 4 is symmetric with respect to a sign change in $\theta$ and an increase in phase $\omega_o t$ by $\pi$ radians, and the attractor for small drive amplitudes $g \ll 1$ has a corresponding symmetry. As $g$ increases, symmetry breaking[32] occurs, and symmetry-related pairs of attractors are formed. For the conditions $g > 1$ that we consider below, the pendulum can swing over the top, and pairs of running modes with positive and negative nonzero average angular velocity $\langle d\theta/dt \rangle$ occur. These modes correspond to the zero current voltage steps in an r.f.

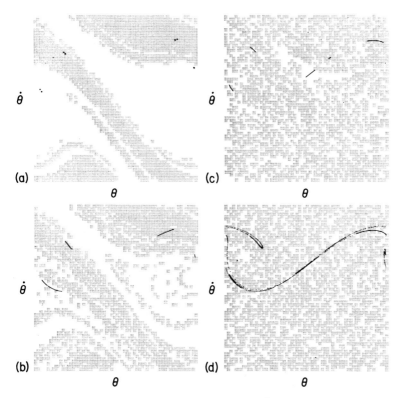

Fig. 8. Basins of attraction in angle $\theta$ (range 0 to $2\pi$) and angular velocity $d\theta/dt$ (range -3 to 3) for positive (shaded) and negative (blank) running modes computed for $\omega_O = 2/3$, Q = 2, and drive amplitudes (a) g = 1.4600, (b) g = 1.4800, (c) g = 1.4954, and (d) g = 1.4955. Poincare sections of the attractors are shown at drive phase $\omega_O = 2\pi$ (Ref. 27).

driven Josephson junction. The choice between positive and negative running modes is determined by the initial conditions on $\theta$ and $d\theta/dt$.

Two examples of the development with increasing drive of Poincare sections of the attractors for positive and negative running modes and their associated basins of attraction are shown in Figs. 8a to 8d, taken from Ref. 27. The set of initial conditions which yield positive running modes is shaded, and the set which yields negative modes is left blank. In Figs. 8a and 8b the Poincaré sections show a transition from two period 4 running modes to two chaotic running modes which remain distinct and do not show intermittency. A crisis[14-16] occurs between Figs. 8c and 8d when the pair of banded chaotic attractors shown in Fig. 8c intersect the basin boundary and lose their individual stability. The motion on the single attractor thus formed, shown in Fig. 8d, is intrinsically intermittent just above the crisis: the time-averaged angular velocity switches between positive and negative values on a time scale much longer than one cycle of the resonant frequency $\omega_o$.

The most striking feature of Fig. 8 is the complexity of the basins of attraction for the positive and negative running modes. At a relatively low drive amplitude for which the attractor is simple and periodic, the basin boundary has a fractal dimension[22] $d > 1$, as illustrated in Fig. 8a where $d = 1.63$. As the drive amplitude increases in Figs. 8b to 8c, just below the crisis, the boundary dimension increases to $d = 1.97$, for which the boundary covers nearly all of phase space. The fact that the basin boundary can be fractal with dimension $d \cong 2$ implies that the final state of the system can be sensitive to even a small uncertainty $\varepsilon$ in the initial conditions, and suggests exceptional sensitivity to external noise, a topic considered below. Grebogi et al. have calculated[17] the scaling of the fraction f of uncertain final states with $\varepsilon$ and find $f \propto \varepsilon^{2-d}$. For dimensions $d \cong 2$ this relation implies that the final state uncertainty is practically independent of $\varepsilon$.

Crisis-induced intermittency occurs as transitions between two or more chaotic attractors, and is qualitatively different from Pomeau-Manneville intermittency[12], which occurs as the loss of stability of a periodic attractor.

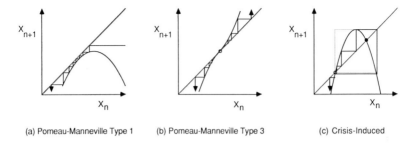

(a) Pomeau-Manneville Type 1     (b) Pomeau-Manneville Type 3          (c) Crisis-Induced

Fig. 9. Schematic illustration of three types of intermittency using a parabolic map as indicated.

Figure 9 illustrates this difference using the example of a
parabolic one-dimensional map. As shown in Figs. 9a and 9b,
Pomeau-Manneville type 1 intermittency occurs just below a
saddle-node bifurcation as the orbit is almost trapped near
the eventual position of the pair of fixed points, and type
3 intermittency occurs as the orbit slowly escapes from a
marginally unstable fixed point[12]. By contrast,
crisis-induced intermittency occurs when the chaotic orbit
intersects the boundary of its basin of attraction[16],
indicated by the dashed square in Fig. 9c.

Angular velocity power spectra[27] for intrinsic
crisis-induced intermittency in the damped driven pendulum
are shown in Figs. 10b and 10c, with a portrait of the
intermittent attractor in Fig. 10a. Typically, the power
spectra are Lorentzians as shown in Fig. 10c, as one would
expect from a simple hopping model with a single
characteristic time; in Fig.10c the characteristic hopping
time is $\cong$ 10 drive cycles. However, just above the crisis
the power spectrum is approximately 1/f over more than two
decades in frequency. We have also found 1/f noise for
other parameter combinations, and it is not uncommmon. The
source of 1/f noise in this system is not clear, although it
occurs near parameter values for which the basins of
attraction are highly complex, as shown in Fig.8c.

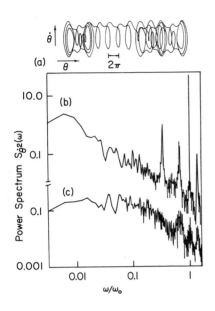

Fig. 10. (a) Phase portrait, and (b), (c) angular velocity power spectra computed from Eq. 4 without added noise for $\omega_0 = 2/3$, $Q = 2$; (a) and (c) g = 1.5000; (b) g = 1.4955 (from Ref. 27).

As discussed above, external noise can induce intermittency in systems with multiple attractors which are intrinsically stable. The noise-induced hopping rate is determined by two factors: the expansion of the attractor with noise and the proximity and complexity of the basin boundary. Figures 11a and 11b compare Poincaré sections of the intrinsically intermittent attractor in Fig. 11a in the absence of noise and the noise-broadened attractor at a lower drive amplitude for which the intrisic orbit was simply periodic[28]. As shown, the influence of external noise is to cause expansion of the periodic attractor in Fig. 11b along directions for which phase space contraction is weakest, so that it strongly resembles the intrinsically intermittent attractor in Fig. 11a. When the system is chaotic, these directions correspond to phase space expansion with a positive Lyapunov exponent. The striking similarity of the noise-broadened and intrinsically

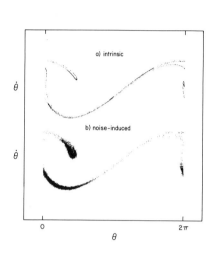

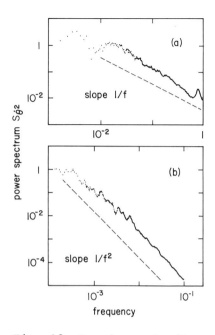

Fig. 11. (a) Intrinsically
chaotic unlocked attractor
without added noise, and (b)
noise-broadened periodic
attractor.  Parameter values:
$\omega_0$=2/3, Q=2,  (a) g=1.4955,
(b) g=1.3000 (from Ref.28).

Fig. 12. Angular velocity
power spectra for noise-
induced intermittency;
$\omega_0$=2/3, Q=2, (a) g=1.4954,
and (b) g=1.3000 (from
Ref. 28).

intermittent attractors in Figs. 11b and 11a is typical, and
is apparently a consequence of the pattern of flow in phase
space induced by Eq. (4).

Power spectra for noise-induced intermittency are shown
in Fig. 12 at two drive amplitudes[28].  Typically, the
noise-induced power spectrum is Lorentzian as in Fig. 12b,
as expected for the simplest models.  Despite its
simplicity, this type of noise can be very difficult to deal
with experimentally, because the characteristic hopping time
can be arbitrarily long; in Fig. 12b computed with a
relatively large noise amplitude, the hopping time is

$10^3$ drive cycles. An approximately 1/f power spectrum is found for parameter values just below the crisis in Fig. 8, where the basins of attraction are very complex, as shown in Fig. 12a.

Chaos-induced intrinsic intermittency and noise-induced extrinsic intermittency have surpringly similar characteristics. Both the shape of the attractor and the resulting power spectrum are very similar, and these two cases may prove difficult to distinguish experimentally. From our digital simulations where the noise level is controlled, we find that the nature of the motion is determined by whichever mechanism, chaos or noise induced fluctuations, produces the largest excursions in the attractor. Fractal basin boundaries also tend to favor noise-induced intermittency. Recently Iansiti et. al.[40] have established a correlation between the experimentally measured noise power of intermittent hopping between voltage steps in an r.f. driven Josephson junction and the computed basin boundary dimension.

We briefly consider another example where the concepts of multiple attractors, basin boundaries, and intermittency may prove important. This is the realization of a content addressable memory (CAM) as a nonlinear electronic network of threshold switching elements, recently discussed by Hopfield[41,42] and others[43-45]. These circuits are analogs of neural networks in biological systems and they show considerable promise for applications in fast optimization and pattern-recognition problems. One way of constructing such a network is to interconnect many identical inverting-noninverting amplifier pairs via a matrix of resistors, so that each input can be coupled to the outputs of every amplifier, as described by Hopfield[41]. The

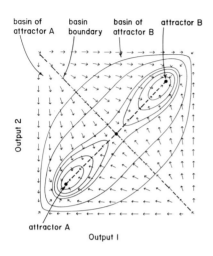

basin of attractor A   basin boundary   basin of attractor B   attractor B

Output 2

attractor A

Output 1

Fig. 13. Attractors, basins of attraction, and flow for a simple example of a content-addressable memory taken from Ref. 41.

resulting equations of motion for the network can be written as:[41]

$$C_i(du_i/dt) = \sum_j T_{ij}V_j - u_i/R_i + I_i,\qquad(5)$$

$$V_i = g_i(u_i);$$

where $u_i$ and $V_i$ are the inputs and outputs, g is the nonlinear transfer function, $T_{ij}$ is the coupling matrix, and $R_i$ and $C_i$ are the resistance and capacitance shunting the input of each amplifier. Proper selection of the conductance matrix $T_{ij}$ creates multiple stable states for the network, which represent stored patterns, or "memories". For successful pattern recognition, for example, an arbitrary initial condition $V_{ij}$ representing a noisy image would relax to the stable state pattern which it most resembles.

Figure 13 is a modified version of a simple example given by Hopfield[41], which acts as a flip-flop. The flow with time of an initial condition, shown in Fig. 13, tends toward the nearest final state, A or B, as discussed above.

To emphasize that this network is a nonlinear dynamical system with properties similar to those discussed above, and with possibly complex behavior in general cases, we have labelled the attractors, basins of attraction, and the basin boundary in Fig. 13. For this case the basins of attraction are smooth and well behaved; however in more general cases they may be complex or even fractal. Smooth boundaries at predictable locations are essential to the successful operation of the proposed networks, yet very little is presently known about this subject. Other open questions are the response to external drive, and means of loading new initial conditions. For example, the basins of attraction for the damped driven pendulum are simple and well behaved without drive; with periodic drive they can evolve into complex structures such as those shown in Fig. 8 above.

Intermittency in nonlinear dynamical systems is a relatively new subject, and many theoretical issues and experimental systems need to be investigated. Theoretical subjects for study include other possible sources of intrinsic intermittency, the statistics of the motion of orbits on chaotic attractors and on noise broadened attractors, the relative contributions of intrinsic and noise-induced intermittency in physical systems, and the origins of $1/f$ noise in chaotic systems. Intermittency can be important in a broad range of experimental systems including extrinsic photoconductors, frequency mixers, Josephson junctions, physical pendula, and content addressable memories. For experimentalists, intermittency is particularly challenging and fascinating, because intermittent phenomena are not usually thought to arise from the fundamental equations of motion of a device.

## 4. Acknowledgements

We thank E. E. Haller, C. D. Jeffries, M. R. Beasley, M. Tinkham, E. Ott, J. A. Yorke, and B. Mandelbrot for helpful discussions. This work was supported by the Office of Naval Research under contracts N00014-84-K-0329 and N00014-84-K-0465. One of us (E.G.G.) acknowledges support as an AT&T Bell Laboratories Scholar.

## References

1. J. Guckenheimer and P. Holmes, Nonlinear Oscillations, Dynamical Systems, and Bifurcations of Vector Fields (Springer-Verlag, New York, 1983), and references therein.
2. N. Minorsky, Nonlinear Oscillations (Krieger, Malabar, Florida, 1983), and references therein.
3. S. W. Teitsworth, R. W. Westervelt, and E. E. Haller, Phys. Rev. Lett. 57, 825 (1983).
4. J. Peinke, A. Muhlbach, R. P. Huebener, and J. Parisi, Phys. Lett. 108A, 407 (1985).
5. E. Aoki and K. Yamamoto, Phys. Lett. 98A, 72 (1983).
6. D. C. Seiler, C. L. Littler, and R. J. Justice, Phys. Lett. 108A, 462 (1985).
7. P. R. Bratt, in Semiconductors and Semimetals, vol. 12, edited by R. K. Willardson and A. C. Beer (Academic, New York, 1977), p. 39.
8. S. W. Teitsworth and R. M. Westervelt, in Proc. 8th Int. Conf. on Noise in Physical Systems, Rome, 1985, to be published.
9. S. W, Teitsworth and R. M. Westervelt, to be published.
10. S. W. Teitsworth and R. M. Westervelt, Phys. Rev. Lett. 53, 2587 (1984).
11. R. M. Westervelt and S. W. Teitsworth, J. Appl. Phys. 57, 5457 (1985).
12. Y. Pomeau and P. Manneville, Commun. Math. Phys. 74, 189 (1980).
13. P. Manneville, J. Physique (Paris) 41, 1235 (1980).
14. C. Grebogi, E. Ott, and J. A. Yorke, Phys. Rev. Lett. 48, 1507 (1982).
15. C. Grebogi, E. Ott, and J. A. Yorke, Phys, Rev. Lett. 50, 935 (1983).
16. C. Grebogi, E. Ott, and J. A. Yorke, Physica (Utrecht) 7D, 181 (1983).
17. C. Grebogi, S. W. McDonald, E. Ott, and J. A. Yorke, Phys. Lett. 99A, 415 (1983).
18. F. T. Arecchi, R. Badii, and A. Politi, Phys. Rev. A 29, 1006 (1984); Phys. Rev. A 32, 402 (1985).

19. I. Procaccia and H. Schuster, Phys. Rev. A 28, 1210 (1983).

20. T. Geisel and S. Thomae, Phys. Rev. Lett. 52, 1936 (1984); T. Giesel, J. Nierwetberg and A. Zacherl, Phys. Rev. Lett. 54, 616 (1985).

21. G. Mayer-Kress and H. Haken, Physica (Utrecht) 10D, 329 (1984).

22. B. B. Mandelbrot, The Fractal Geometry of Nature (Freeman, San Francisco, 1983), and references therein.

23. S. W. McDonald, C. Grebogi, E. Ott, and J. A. Yorke, "Fractal Basin Boundaries," plasma preprint UMLPF 84-017, and references therein.

24. H. Hartnagel, Semiconductor Plasma Instabilities (American Elsevier, New York, 1969).

25. S. M. Sze, Physics of Semiconductor Devices (Wiley, New York, 1981).

26. E. G. Gwinn and R. M. Westervelt, in Proceedings of the Seventeenth International Conference on Low-Temperature Physics, Karlsruhe, West Germany, 1984, edited by U. Eckern, A. Schmid, W. Weber, and H. Wuhl (North-Holland, Amsterdam, 1984) p.1139.

27. E. G. Gwinn and R. M. Westervelt, Phys. Rev. Lett. 54, 1613 (1985).

28. E. G. Gwinn and R. M. Westervelt, to be published.

29. B. A. Huberman, J. P. Crutchfield, and N. H. Packard, Appl. Phys. Lett. 37, 750 (1980).

30. R. L. Kautz, J. Appl. Phys. 52, 3528, 6241 (1981).

31. N. F. Pederson and A. Davidson, Appl. Phys. Lett. 39, 830 (1981).

32. D. D'Humieres, M. R. Beasley, B. A. Huberman, and A. Libchaber, Phys. Rev. A 26, 3483 (1982).

33. M. Octavio, Phys. Rev. B 29, 1231 (1984).

34. I. Goldhirsch, Y. Imry, G. Wasserman, and E. Ben-Jacob, Phys. Rev. B 29, 1218 (1984).

35. R. F. Miracky, M. H. Devoret, and J. Clarke, Phys. Rev. A 31, 2509 (1985).

36. R. F. Miracky, J. Clarke, and R. H. Koch, Phys. Rev. Lett. 50, 856 (1983).

37. V. N. Gubankov, K. I. Konstantinyan, V. P. Koshelets, and G. A. Ovsyannikov, IEEE Trans. Magn. 19, 637 (1983).

38. M. Octavio and C. Readi Nasser, Phys. Rev. B 30, 1586 (1984).

39. Q. Hu, J. U. Free, M. Iansiti, O. Liengme, and M. Tinkham, in Proc. Conf. Applied Superconductivity, 1984 (to be published).

40. M. Iansiti, Q. Hu, R. M. Westervelt, and M. Tinkham, Phys. Rev. Lett. 55, 746 (1985).

41. J. J. Hopfield, Proc. Natl. Acad. Sci. U.S.A. 81, 3088 (1984).

42. J. J. Hopfield and D. W. Tank, "Neural Computation of Decisions in Optimization Problems", Biological Cybernetics, to appear.

43. T. Kohonen, <u>Content Addressable Memories</u> (Springer-Verlag, New York, 1980), and references therein.

44. Choi and Huberman, Phys. Rev. B **28**, 2547 (1983).

45. S. Wolfram, Rev. Mod. Phys. **55**, 601 (1985).

NONLINEAR DYNAMICS OF SOLID STATE SYSTEMS

Carson D. Jeffries
Department of Physics
University of California
Berkeley, CA   94720

## ABSTRACT

A short review is given of a number of phenomena
in solid state physics which display nonlinear
dynamics.  Two examples are discussed in more
detail:  plasma waves in germanium crystals and
spin waves in ferrites.  Solids are often well
characterized and have diverse properties; they
are quite interesting from the viewpoint of
experimental and theoretical nonlinear dynamicists.
A fundamental understanding of their dynamics will
have significant bearing on solid state device
technology and applications.

1.   Introduction:  Perspectives

From my viewpoint as an experimentalist, these are some of the outstanding issues in nonlinear dynamics:

A.   Identification of universal features:  How well and under
what conditions can the overall temporal behavior of a real nonlinear
system be viewed as *composed of elementary recognizable elements* (e.g.,
period doubling, quasiperiodicity, entrainment, chaos, intermittency,
specific power s p e c t r a, ...), characterized by scaling relations
and universal numbers computed from elementary models, usually maps.
How useful is this approach?

B.   Spatio-temporal behavior:  This is the most general problem
and requires simultaneous measurement of temporal behavior at a large
number of spatial elements of an extended system, e.g., a real fluid or

a plasma. Of special interest is the transition to weak turbulence
(chaos) and then to strong turbulence. Are there optimum data-taking
schemes and analysis procedures? How can spatio-temporal chaos be
characterized? What to do when the fractal dimension becomes intractably
large?

C. Evolution of structures in complex systems. I refer here to
relatively slow evolution such as dendritic crystal growth; development
of biological structures; and emergent properties of extended collective
systems, e.g., multidimensional arrays of nonlinear elements.

The above issues are, of course, very general. I focus now on the
large field of solid state physics and list, in Section 2, some specific physical systems of interest. Westervelt in his contribution to
this volume presents an extensive list of nonlinear phenomena in semiconductors.

2. Some Solid State Systems

Perhaps the simplest model of a solid is a set of coupled nonlinear
oscillators. If the dissipation is large enough, these may be modelled
by a set of coupled maps. The next level of modelling might be a set
of nonlinearly coupled modes, roughly applicable to plasma waves, spin
waves, and acoustic waves in crystals. As a simple example consider
three modes of waves of amplitude $C_1$, $C_2$, $C_3$, bilinearly coupled:

$$\frac{dC_1}{dt} = -\gamma_1 C_1 + M_1 C_2 C_3 , \tag{1a}$$

$$\frac{dC_2}{dt} = -\gamma_2 C_2 + M_2 C_1 C_3^* , \tag{1b}$$

$$\frac{dC_3}{dt} = -\gamma_3 C_3 + M_3 C_1 C_2^* , \tag{1c}$$

where $\gamma_i$ and $M_i$ are damping and coupling constants, respectively.
Numerical computations have shown that such equations can have a
Poincare section that reduces to a one-dimensional map and hence display
a period doubling cascade to chaos [1].

We now list some solid state systems that i) have been studied successfully from the viewpoint of contemporary nonlinear dynamics of the last five years; or ii) have been reported earlier to display various instabilities, not really understood, but that can now be fruitfully reexamined, experimentally and theoretically; or iii) systems that will probably display interesting nonlinear dynamics.

A. <u>Plasmas</u>: Helical electron-hole plasma density waves in Ge rods show period doubling, quasiperiodicity, loss of spatial coherence [2-4]. Electro-acoustic interactions in GaAs show subharmonic generation [5]. Other good candidates for study appear to be helicon and Alfvén waves [6]; the two-stream instability [7]; and the magnetic pinch effect [8]. See general references [9-12].

B. <u>Spin waves</u>: Period doubling cascade to chaos observed in spin waves in yttrium iron garnet spheres excited by ferromagnetic resonance [13]. Similar phenomena observed by parallel pumping [14, 15]. Route to chaos by "irregular periods" observed by parallel pumping ferromagnets [16]. General references [17-19].

C. <u>Charge density waves</u>: Materials such as $TaS_3$, $NbSe_3$, driven by ac or dc currents display period doubling, chaos, and quasi-periodicity [20-22].

D. <u>Acoustic waves</u>: Strongly driven Rochelle crystals show a period cascade to chaos at a temperature near the ferroelectric transition [23].

E. <u>Oscillatory conduction in semiconductors</u>: Low temperature photoconductivity studies of pure Ge crystals show period doubling, chaos, and quasiperiodicity [24]. Qualitatively similar behavior is found in GaAs [25]. In the post-breakdown regime in p-Ge, spontaneous oscillations and chaos are found [26]. Oscillatory and chaotic states are found in the conductivity of barium sodium niobate at temperatures ~600°C [27]. See [28] for general references to older experimental work on semiconductor instabilities before the development of contemporary nonlinear dynamics theory; many of these results could be

reexamined and now understood. An understanding of semiconductor
instabilities has important applications in present technology.

F. <u>Josephson junctions</u>: Intermittent chaos is observed in
resistively shunted Josephson junctions [29]. Much simulation work
has also been reported, including the effects of fractal boundaries of
the basins of attraction on the low frequency power spectra [30].

G. <u>Discrete nonlinear solid state oscillators</u>: Driven resonators
composed of a linear inductance and the nonlinear charge storage
properties of p-n junctions have been extensively studied, and in the
simplest cases of large damping display a period doubling cascade to
chaos and periodic windows [31]. If driven harder, and with less
damping, the system displays more complex behavior due to the extreme
asymmetry pf the effective restoring force:  a period adding sequence
[32-34]. Intermittency [35], effects of added noise [36], and crises
of the attractor [37] have also been studied. If two or more resonators
are coupled, the system displays a Hopf bifurcation to quasiperiodicity,
entrainment horns, and breakup of the invariant torus [34]. This simple
real physical system displays much of the behavior of complex driven
passive nonlinear systems.

Studies of a forced symmetric nonlinear self-oscillator (using a
saturable inductor with hysteresis) show a rich behavior:  symmetry
breaking, quasiperiodicity, entrainment horns, and homoclinic bifurcations [38]. Through direct observation of both stable and unstable
manifolds, the behavior near points of strong and weak resonance is
found to correspond with V. I. Arnold's theory of versal deformations
of the plane [39].

3. Chaos and Turbulence in an Electron-Hole Plasma in a Ge Crystal

As an example of nonlinear dynamics in a solid, we review the
experiments of Held <u>et al</u>. [2-4] at Berkeley on the spatial and temporal behavior of chaotic instabilities of an electron-hole plasma in
a germanium rod.  The plasma is produced by injecting both electrons
and holes into a rod-shaped crystal of germanium at liquid nitrogen
temperatures; the crystal is placed in a magnetic field $B_0$ parallel to

its axis, and an adjustable electric field is also applied along the length of the sample. The plasma can absorb energy from the applied fields and, beyond some threshold (typically a few volts/cm at a few kilogauss), an unstable travelling helical density wave develops within the plasma. Several nonlinearly coupled modes can be excited within the boundaries of the crystal.

Experimentally we measure the total current $I(t)$ through the crystal and the potential across it, $V(t)$, as the driving parameter $V_{dc}$ is increased. By also recording the voltages $V_i(t)$ across pairs of probe contacts formed along the length of the sample, we can observe spatial variations in the plasma density. At the onset of the helical insta- bility, spontaneous current oscillations are observed. As $V_{dc}$ is increased further, we find that this simple physical system exhibits complex nonlinear dynamics, including a period doubling route to chaos when only one mode is excited. More generally, when more modes are excited, we observe quasiperiodicity; self-entrainment; temporal chaos; and a partial loss of spatial coherence -- indicating the spatial break- down of the helical density wave and the onset of "turbulence" in this solid state system.

Our experiments are, of course, related to some hydrodynamic experi- ments on fluids, e.g., Rayleigh-Bénard convection and Couette-Taylor flows, as well as other experiments on nonlinear dynamical systems. Such experiments are partly motivated by the conjecture that in dissi- pative nonlinear media the dynamics may be modelled by a strange attractor of relatively low dimension, in contrast to a very large number of degrees of freedom associated with ergodic systems. Coherent oscillations of the type we study were originally observed by Ivanov and Ryvkin [40] in Ge and were subsequently studied both theoretically [41,42] and experimentally [42,43] in a number of other semiconductors. It is possible that chaotic behavior was observed earlier but not recognized as such, owing to the lack of mathematical framework now available. Our physical system is well characterized, and the equations of motion well known [2]. In the simplest case the equations can be approximated by Eq. (1). This appears to be a good system for detailed study of spatio-temporal plasma turbulence and, in fact, it is the first plasma system found to exhibit the universal period doubling and quasiperiodic transitions to chaos.

Perhaps the single feature most useful in characterizing the plasma is the power spectrum of the current, $|I(\omega)|^2$, from which we can detect the onsets of spontaneous oscillations, period doubling, quasiperiodicity, and chaos. However, observation of only power spectra does not enable us to distinguish between deterministic chaos and stochastic noise; both result in broadened spectral peaks. To uniquely identify the observed spectral broadening as deterministic chaos, we observe in real time the phase portrait, a plot of $V(t)$ vs. $I(t)$; and the first return map, a plot of $I_n$ vs. $I_{n+1}$, where $\{I_n\}$ is the set of local current maxima. The return map is topologically equivalent to a Poincaré section of the attractor. When the return map does not fill an entire area within 2-dimensional space, the motion of the system is confined to a low-dimensional strange attractor. However, a system in which the return map does fill an entire area within 2-dimensional space may still be characterized by low-dimensional chaos (with attractor dimension typically $\geq 2.5$). In these cases even a return map cannot distinguish between chaos and stochastic noise, and one must consider more quantitative measures of the dimensionality of the system.

The fractal dimension [44] provides just such a quantitative measure and thus an approximate measure of the number of degrees of freedom needed to characterize the plasma at any instant of time. We use the following procedures [45] to measure the fractal dimension d of our plasma instabilities: we begin by recording a data set of N values of the current at uniformly spaced time intervals [i.e., $I(t+mT) \rightarrow I(mT)$, $m = 1,2,...,N$] using a fast 12-bit analog-to-digital converter and an LSI-11/23 computer. From the data set $\{I(T), I(2T), ..., I(NT)\}$ we construct $N - D + 1$ vectors $\vec{G}_m = [I(mT), I((m+1)T), ..., I((m+D-1)T)]$ in a D-dimensional phase space; D is referred to as the embedding dimension of the reconstructed phase space $\vec{G}$. Next, we compute the number of points on the attractor, $N(\varepsilon)$, which are contained within a D-dimensional hypersphere of radius $\varepsilon$ centered on a randomly selected vector $\vec{G}_m$.

One expects scaling of the form

$$N(\epsilon) \propto \epsilon^d ,$$ (2)

where d is the fractal dimension of the attractor. Thus, a plot of
log $\overline{N(\epsilon)}$ vs. log $\epsilon$ is expected to have a slope d, where $\overline{N(\epsilon)}$ is the
average for hyperspheres centered on many different $\vec{G}_m$. This procedure
is carried out for consecutive values of D = 2, 3, 4, ..., until the
slope has converged. This is done to ensure that the embedding dimension
chosen is sufficiently large (important if the dimension of the phase
space is not known), and to discriminate against high dimensional stochastic noise, not of deterministic origin.

To determine whether or not a plasma density wave is spatially
coherent, we compare the fluctuations in plasma density at different
points along the sample. We obtain a crude measure of the degree of
coherence by using a fast two-channel digital storage oscilloscope and
comparing the voltages $V_i(t)$ across pairs of contacts located at different
positions along the z-axis of the sample. If the temporal behavior of
I(t) is periodic, we observe only a phase shift in $V_i(t)$ along z, which
indicates a coherent travelling wave.

To obtain a more quantitative measure of the degree of spatial
coherence, applicable for nonperiodic behavior, we calculate a spatial
correlation function C(r), defined as

$$C(r) = \frac{2}{N} \left| \sum_{n=1}^{N} V_i(nT)V_j(nT) \right| ,$$ (3)

where $V_i(t)$ and $V_j(t)$ are the voltages across two pairs of contacts
separated by a distance r, T is the sampling time interval, and N is a
number large enough that C(r) has converged, typically 20,000. We find
that C(r) is independent of T.

Results: Temporal routes to chaos. In different regions of parameter
space we observe different types of transitions to chaos. A sequence was
taken with $B_0$ = 4 kG, as $V_{dc}$ was increased from 0 to 25 V. The overall
behavior of I(t) was found to be as follows: For $V_{dc}$ < 6 V, I(t) has

only a dc component. At $V_{dc}$ = 6 V, $I(t)$ spontaneously becomes periodic. Regions of chaotic dynamics occur in the intervals $7.0 \leq V_{dc} \leq 7.4$ V; $10.0 \leq V_{dc} \leq 10.7$ V; and $14.9 \leq V_{dc} \leq 18$ V; otherwise, $I(t)$ is periodic. The clearest of these three chaotic sequences starts at $V_{dc}$ = 10.0 V: $I(t)$ is oscillating at a fundamental frequency $f_0 \approx$ 118 kHz, i.e., at period 1. The phase portrait, $I(t)$ vs. $V(t)$, shows that the oscillation has a small spectral component at a harmonic of $f_0$. However, there is no subharmonic component. As $V_{dc}$ is increased, $I(t)$ shows a period doubling bifurcation: the emergence of a spectral component at $f_0/2$. At larger $V_{dc}$, another period-doubling bifurcation occurs with new spectral components at $f_0/4$, $3f_0/4$, $5f_0/4$, ... . At slightly larger $V_{dc}$ $I(t)$ becomes nonperiodic and its power spectrum enters a region of broadband "noise". For further increases of $V_{dc}$ there appear noise-free windows of periods 3, 4, 5, ..., within this region of broadband noise. This sequence ends at $V_{dc}$ = 10.7 V with a return to period 1 oscillations.

A second type of transition which we have observed is the quasi-periodic route to chaos: as $V_{dc}$ is increased, the onset of a quasi-periodic state is followed by a transition to chaos. In one such sequence taken at $B_0$ = 11.15 kG, at $V_{dc}$ = 2.865 V, $I(t)$ is spontaneously oscillating at a fundamental frequency $f_1$ = 63.4 kHz. At $V_{dc}$ = 2.907 V, the system becomes quasiperiodic: a second spectral component appears at $f_2$ = 14 kHz, incommensurate with $f_1$. At $V_{dc}$ = 2.942 V, the system is still quasiperiodic; however, the two modes are interacting and the nonlinear mixing gives spectral peaks at the combination frequencies $f = mf_1 + nf_2$, with m,n positive and negative integers. As $V_{dc}$ is increased further, we observe a series of frequency lockings, i.e., $(f_1/f_2)$ = rational number, until the onset of chaos is reached, indicated by a slight broadening of the spectral peaks. As $V_{dc}$ is increased further, the spectra become even broader. This is followed by a return to quasiperiodicity at $V_{dc}$ = 3.125 V and, subsequently, simple peri-odicity at $V_{dc}$ = 3.442 V. For $V_{dc}$ = 3.058 V we measure the fractal dimension of the attractor, as discussed above, finding $d \approx 2.6$; con-vergence is obtained both with respect to the embedding dimension D and the number of data points ($N \approx 10^4$).

Results: Spatial behavior. Turning attention to the question of spatial coherence within the instabilities, we ask whether the chaotic states we observe correspond to a temporally chaotic yet still spatially coherent or whether the onset of chaos corresponds to a breakup of spatial order within the density wave.

For our system we define a transition to "weak" turbulence to be one in which the transition from periodicity to chaos is followed by a transition back to periodicity as $V_{dc}$ is increased further. The two scenarios discussed above both correspond to transitions to "weak" turbulence. For this case (data taken at $B_0 = 11.15$ V, $V_{dc} \approx 5$ to 6 V) we calculate the correlation function $C(r)$, Eq. (3), for the periodic state and find that it is fit by the correlation function for a travelling wave - not surprising. In addition, we find that the quasiperiodic and chaotic states both have correlation functions that follow the periodic case and so conclude that this weakly turbulent instability is chaotic in the temporal domain only. Even while exhibiting chaotic behavior it remains essentially a spatially coherent plasma density wave.

However, with sufficiently large applied electric and magnetic fields, we find we can drive the plasma into a turbulent state which will not become periodic again. Instead, all of the frequency peaks in the power spectrum merge into a single, broad, noiselike band. We classify this as a transition to "strong" turbulence. An example is found at $B_0 = 11.15$ kg as $V_{dc}$ is increased from 10.0 V to 21.8 V. At 11.6 V the system becomes quasiperiodic, and chaotic at 12.1 V, with measured fractal dimension $d = 2.6$. At 12.9 V the measured fractal dimension has increased to $d \geq 8$: the fractal dimension plots do not show a convergence of slope for embedding dimensions as large as $D = 18$ and number of data points $N = 884,000$. Thus we can only set a lower limit to the value of $d$. At 13.8 V the power spectra are broad with a few peaks, and at 21.8 V, very broad with no peaks.

This difficulty in calculating large fractal dimensions is a problem encountered whenever one works with a very chaotic system; the number of data points required for convergence increases exponentially with the fractal dimensions of the system. At present, although we know that our system experiences a large jump in dimensionality, we have not yet determined whether this onset is characterized by chaotic dynamics of an attractor of fractal dimension many orders smaller than the number of degrees of freedom of the particles in the system. In the same scenarios which shows the jump in dimension we also find a gradual loss in spatial coherence, as observed from the oscilloscope voltage traces between pairs of probe contacts and the measured correlation function $C(r)$, which decreases with increasing $V_{dc}$.

4.    Chaotic Dynamics of Spin Wave Instabilities in Ferrites

Noisy instabilities in ferromagnetic resonance saturation in some ferrites were experimentally discovered in the 1950's [46] and explained by Suhl [17] in a detailed theory of nonlinear coupling between the uniform precession mode of the magnetization vector and spin waves. The uniform mode can excite the spin waves which grow exponentially. Suhl recognized early that "...This situation bears a certain resemblance to the turbulent state in fluid dynamics...". In the simplest case the equations are similar to Eqs. (1), where $C_1$ is the amplitude of the uniform mode (wave vector k = 0); $C_2$ and $C_3$ are amplitudes of a pair of spin waves (wave vector k and -k); and a radio frequency driving term must be added to Eq. (1a), as well as higher order nonlinear terms of the form $C_1 C_2 C_3^*$, the so-called Suhl $2^{nd}$ order terms.

We review below recent experiments at Berkeley on spheres of gallium yttrium iron garnet (YIG) which show that Suhl's $2^{nd}$ order instability is a period doubling cascade to chaos [13]. The sample magnetization is now known to display temporal chaos; it is not yet known if it also displays spatial incoherence as in the plasma case. Parallel pumping experiments on YIG [14] and in copper salts [15] also show period doubling. In some salts a route to chaos through irregular periods without period doubling has been repeated [16]. Suhl's theory has recently been extended [47] by numerical computation, and a period doubling cascade to chaos is found. Similar computations have been carried out for the parallel pumping case [14,48].

The Berkeley experiments are carried out by mounting a highly polished sphere of Ga-YIG in a magnetic field $H_0 \parallel z$; pumping with a field $H_1 \parallel x$ at the ferromagnetic resonance frequency $f_0$; and observing a signal $V_s(t)$ which is the time derivation of the transverse magnetization. $V_s(t)$ has a strong component at $f_0$, which is the usual ferromagnetic resonance signal. In our case $H_s \approx 460$ Oe; $f_0 = 1.3 \times 10^9$ Hz; sample radius $\approx 0.047$ cm and saturation magnetization $4\pi M_s = 300$ gauss; resonance line width $\approx 0.5$ gauss. As the driving field $H_1$ is increased, there is a threshold value $H_{1a}$ at which low frequency self-oscillations set in at $f_{1a} \approx 250$ kHz, corresponding to annihilation of two $(\omega_0, k=0)$ magnons and the creation of a spin wave pair $(\omega_k, k)$ and $(\omega_{k'}, -k)$. The value of the frequency $f_{1a}$ corresponds closely to the lowest standing wave mode in the spherical sample of a packet of such spin waves travelling parallel to $H_0$ [49]. We note that numerical solutions [47,14] of the coupled mode equations predict self-oscillations arising from a Hopf bifurcation to a limit cycle at a frequency determined by the coupling parameters and independent of the sample size. However, the calculated frequency does not yet agree with observation and the question is at present unresolved.

As $H_1$ is increased still further, there is another threshold $H_{1b}$ at which a second self-oscillation sets in at $f_{1b} \approx 16$ kHz, which we interpret as the onset of creation of a spin wave pair with $k \approx 0$. From theory, the relaxation process becomes exponentially weaker as $k \to 0$ [50]; we observe long lifetimes for these low frequency self-oscillations, as well as a period doubling cascade to chaos and periodic windows. This is the first experiment to demonstrate the existence of chaotic dynamics in magnetic materials and much work remains to be done both experimentally and theoretically. Spin systems are well characterized and the macroscopic nonlinear parameters in the coupled mode equations can be calculated microscopically. This feature makes possible, in principle, the comparison between a nonlinear dynamics experiment and high level theory.

## Acknowledgments

I wish to acknowledge, with much thanks, the major contributions to the plasma experiment by G. A. Held and E. E. Haller, and to the spin wave experiment by George Gibson. James Crutchfield, Paul Bryant,

Alan Portis, and Robert M. White gave helpful discussions. This work was supported by the Director, Office of Energy Research, Office of Basic Energy Sciences, Materials Sciences Division of the U.S. Department of Energy under Contract No. DE-AC03-76SF00098.

References

1. J.-M. Wersinger, J. M. Finn, and E. Ott, Phys. Fluids <u>23</u>, 1142 (1980).
2. G. A. Held, Carson Jeffries, and E. E. Haller, Phys. Rev. Lett. <u>52</u>, 1037 (1984).
3. G. A. Held, C. Jeffries, and E. E. Haller, <u>Proc. 17th Int. Conf. Phys. Semicond.</u>, eds. J. D. Chadi and W. A. Harrison (Springer-Verlag, 1985), p. 1289.
4. G. A. Held and Carson Jeffries, Phys. Rev. Lett. <u>55</u>, 887 (1985).
5. M. Schulz and B. K. Ridley, Phys. Lett. <u>29A</u>, 17 (1969).
6. A. Libchaber and R. Veilex, Phys. Rev. <u>127</u>, 774 (1962); R. Bowers, C. Legendy, and R. Rose, Phys. Rev. Lett. <u>7</u>, 339 (1961); C. C. Grimes and S. J. Bucksbaum, Phys. Rev. Lett. <u>12</u>, 357 (1964); G. A. Williams, Phys. Rev. A <u>139</u>, 771 (1965).
7. D. A. Pines and J. R. Schrieffer, Phys. Rev. <u>124</u>, 1387 (1961).
8. M. Glicksman and M. C. Steele, Phys. Rev. Lett. <u>2</u>, 461 (1959); B. Ancker-Johnson and J. E. Drummond, Phys. Rev. <u>131</u>, 1961 (1963).
9. Hans Hartnagel, <u>Semiconductor Plasma Instabilities</u> (Elsevier Publishing Co., New York, 1969).
10. P. M. Platzman and P. A. Wolff, <u>Waves and Interactions in Solid State Plasmas</u> (Academic Press, New York, 1973).
11. J. Weiland and H. Wilhelmssen, <u>Coherent Nonlinear Interaction of Waves in Plasmas</u> (Pergammon Press, New York, 1977).
12. R. Z. Sagdeev and A. A. Galeev, <u>Nonlinear Plasma Theory</u> (Benjamin, New York, 1969).
13. George Gibson and Carson Jeffries, Phys. Rev. A <u>29</u>, 811 (1984).
14. S. M. Rezende, F. M. de Aguiar and O. F. de Alcantara Bonfim, <u>Proc. International Conference on Magnetism</u>, San Francisco, 1985 (in press).
15. Hitoshi Yamazaki, J. Phys. Soc. Japan <u>53</u>, 1153 (1984).
16. F. Waldner, D. R. Barberis, and H. Yamazaki, Phys. Rev. A <u>31</u>, 420 (1985).
17. H. Suhl, J. Phys. Chem. Solids <u>1</u>, 209 (1957).
18. Marshall Sparks, <u>Ferromagnetic Relaxation Theory</u> (McGraw Hill, New York, 1964).
19. Benjamin Lax and Kenneth J. Button, <u>Microwave Ferrites and Ferrimagnetics</u> (McGraw Hill, New York, 1962).
20. G. Gruner and A. Zettl, Phys. Reports, 1985 (in press).
21. S. E. Brown, G. Mozurkewich, and G. Gruner, Phys. Rev. Lett. <u>54</u>, 2272 (1984).
22. R. P. Hall, M. Sherwin, and A. Zettl, Phys. Rev. B <u>29</u>, 7076 (1984); M. Sherwin, R. Hall, and A. Zettl, Phys. Rev. Lett. <u>53</u>, 1387 (1984).
23. David J. Jefferies, Phys. Lett. <u>90A</u>, 316 (1982).

24. W. Teitsworth, R. M. Westervelt, and E. E. Haller, Phys. Rev. Lett. $\underline{51}$, 825 (1983); S. W. Teitsworth and R. M. Westervelt, Phys. Rev. Lett. $\underline{53}$, 2587 (1984).

25. E. Aoki and K. Yamamoto, Phys. Lett. $\underline{98A}$, 72 (1983).

26. J. Peinke, A. Muhlbau, R. P. Huebener, and J. Parisi, Phys. Lett. $\underline{108A}$, 407 (1985).

27. S. Martin, H. Leber, and W. Martiennsson, Phys. Rev. Lett. $\underline{53}$, 303 (1984).

28. Symposium on Instabilities in Semiconductors, IBM J. Res. Develop. $\underline{13}$, 485-644 (1969).

29. Robert F. Miracky, John Clarke, and Roger H. Koch, Phys. Rev. Lett. $\underline{50}$, 851 (1983); Robert F. Miracky, Ph.D. Thesis, 1984, Dept. of Physics, Univ. of California, Berkeley (unpublished) (Lawrence Berkeley Laboratory Report No. LBL-18193); C. C. Chi, C. Vanneste, and D. C. Cronemeyer, Bull. Am. Phys. Soc. $\underline{30}$, 421, paper GO-8 (1985).

30. E. G. Gwinn and R. M. Westervelt, Phys. Rev. Lett. $\underline{54}$, 1613 (1985).

31. P. S. Linsay, Phys. Rev. Lett. $\underline{47}$, 1349 (1981); James Testa, José Perez, and Carson Jeffries, Phys. Rev. Lett. $\underline{48}$, 714 (1982).

32. José Pérez , Ph.D. Thesis, 1983, Dept. of Physics, Univ. of California, Berkeley (unpublished) (Lawrence Berkeley Laboratory Report No. LBL-16898); José Pérez , Phys. Rev. A (in press).

33. S. D. Brorson, D. Dewey, and P. S. Linsay, Phys. Rev. A $\underline{28}$, 1201 (1983); Paul S. Linsay, preprint, 1985.

34. Robert Van Buskirk and Carson Jeffries, Phys. Rev. A $\underline{31}$, 3332 (1985).

35. C. Jeffries and J. Pérez , Phys. Rev. A $\underline{26}$, 2117 (1982).

36. J. Perez and C. Jeffries, Phys. Rev. B $\underline{26}$, 3460 (1982); C. Jeffries and K. Wiesenfeld, Phys. Rev. A $\underline{31}$, 1077 (1985).

37. C. Jeffries and J. Pérez , Phys. Rev. A $\underline{27}$, 601 (1983); R. W. Rollins and E. R. Hunt, ibid $\underline{29}$, 3327 (1984); P. S. Linsay, Phys. Lett. $\underline{108A}$, 431 (1985).

38. Paul Bryant and Carson Jeffries, Phys. Rev. Lett. $\underline{53}$, 250 (1984); Paul Bryant and Carson Jeffries, Lawrence Berkeley Laboratory Report No. LBL-16949, January 1984.

39. V. I. Arnold, Geometrical Methods in the Theory of Ordinary Differential Equations (Springer-Verlag, New York, 1983), Sections 21, 34, 35.

40. I. L. Ivanov and S. M. Ryvkin, Zh. Tekh. Fiz. $\underline{28}$, 774 (1958) [Sov. Phys. Tech. Phys. $\underline{3}$, 722 (1958)].

41. R. D. Larrabee and M. C. Steele, J. Appl. Phys. $\underline{31}$, 1519 (1960); R. D. Larrabee, J. Appl. Phys. $\underline{34}$, 880 (1963); J. Bok and R. Veilex, Compt. Rend. $\underline{248}$, 2300 (1958).

42. C. E. Hurwitz and A. L. McWhorter, Phys. Rev. $\underline{134}$, A1033 (1964).

43. M. Glicksman, Phys. Rev. $\underline{124}$, 1655 (1961); M. Shulz, Phys. Status Solidi $\underline{25}$, 521 (1968).

44. See, for example, J. D. Farmer, E. Ott, and J. A. York, Physica (Utrecht) $\underline{7D}$, 153 (1983). We compute a point-wise dimension.

45. This is the method used by A. Brandstäter et al., Phys. Rev. Lett. $\underline{51}$, 1442 (1983); see Peter Grassberger and Itamar Procaccia, Phys. Rev. Lett. $\underline{50}$, 346 (1983).

46. R. W. Damon, Rev. Mod. Phys. $\underline{25}$, 239 (1953); N. Bloembergen and S. Wang, Phys. Rev. $\underline{93}$, 72 (1954).

47. X. Y. Zhang and H. Suhl, preprint, 1985.
48. K. Nakamura, S. Ohta, and K. Kawasaki, J. Phys. C: Solid State Physics $\underline{15}$, L143 (1982).
49. S. Wang, G. Thomas, and Ta-lin Hsu, J. Appl. Phys. $\underline{39}$, 2719 (1968); G. Thomas and G. Komoriya, J. Appl. Phys, $\underline{66}$, 883 (1975).
50. M. Sparks, Phys. Rev. $\underline{160}$, 364 (1967).

# The Structure of Chaotic Behavior
## In a PN Junction Oscillator

Paul S. Linsay
Department of Physics
Massachusetts Institute of Technology
Cambridge, Massachusetts 02139

## ABSTRACT

The use of driven pn junction oscillators has become important in the study of nonlinear phenomena. In this paper I establish that the nonrepetitative motion often observed with these circuits is chaotic motion, and describe a simple differential equation which can be used to interpret the qualitative behavior of the oscillator and understand the periodic-to-chaotic transitions.

The use of pn junctions as nonlinear elements in electronic circuits has become an important tool in the investigation of nonlinear dynamics[1]. They have played an especially important role in the study of the period doubling route to chaos and the verification of Feigenbaum's theory of universal period doubling transitions[2]. As the use of these circuits becomes more widespread for the study of other types of transitions and the general features of nonlinear dynamics, it is important to establish that the irregular behavior observed is genuine chaotic motion and to find simple models which can be used to qualitatively understand the local and global features of the attractor. Chaotic motion can be identified in a variety of ways, both quantitatively and qualitatively, and three of these will be used in this paper. The establishment of a simple model depends on a detailed understanding of the physics of the electronic circuit which can then be simplified to extract the essential features of its time dependent output. The success of the simple model depends on it incorporating a nonlinearity which is a useful approximation of the essential nonlinearity in the full equations of motion, and on the ability to make analytic and numerical calculations with it.

The basic apparatus is quite simple, a series combination of a diode, an inductor, and a resistor driven by a low impedence sine wave generator. The current flowing in the circuit was sampled once per cycle, and the resulting sampled time series, $I_1, I_2, \ldots, I_n, \ldots$, stored by computer on a disk for later processing. The sampling was done at the negative-going zero-crossing of the driving sine wave. This was defined to be 0 ° of phase, while 90 ° of phase occurred when the drive amplitude was at its most negative value. The amplitude of the sine wave was used as the control parameter of the circuit and the dynamics of the system were studied as this parameter was varied. A bifurcation diagram of the circuit made from this data is shown in Figures 1a and 1b. Figure 1a was made as the control parameter increased, and 1b with the same element values but with the control parameter decreasing.

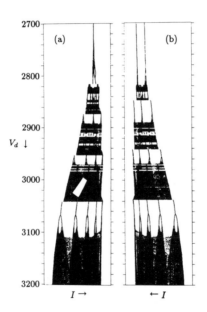

FIG. 1.   Experimental bifurcation diagram with (a) the control parameter, $V_d$, increasing slowly with time, and (b) decreasing slowly with time. Note the reversal of the abscissa in b. Circuit parameters are 1N5470A diode, 25Ω resistor, and 58μH inductor.

The most striking features of these diagrams are the hysteresis which is particularly strong at low drive amplitudes, and the period-adding sequence of the major window, each large window having a period one greater than the previous window as the drive amplitude is increased. The chaotic attractor of the circuit, made by plotting successive current triplets, $\{I_j, I_{j+1}, I_{j+2}\}$, as points in three space, is drawn in Figure 2. (This is actually a Poincaré section of the attractor since the sampled time series is used and not the complete flow.) It is clear from this figure, as well as the preceeding one, that we are not dealing with one-dimensional dynamics, but a higher order dynamical space.

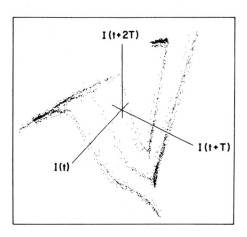

FIG. 2.    The chaotic attractor of Figure 1a at $V_d = 2950$.

A measurement of the dimension of the attractor is the simplest quantitative measurement which can establish that the dynamics of a system is chaotic. A fractional dimension is a good indication that the attractor is heavily folded and thus chaotic. A variety of dimensions and algorithms for their computation have been discussed in the literature[3]. The easiest to implement, and the one used here, is the pointwise dimension. Choose a point, $\{I_j, I_{j+1}, I_{j+2}\}$, at random on the attractor. Let r be the radius of a ball centered on this point, and $N_j(r)$, the number of points of the attractor contained in this ball. If the radius is small enough, the function $N_j(r)$ will have a power law radial dependence,

$$N_j(r) \sim r^{d_j}.$$

The average of the $d_j$ for many randomly chosen points is the pointwise dimension of the attractor, d. The measured value was found to be 1.58 ±0.03, and is largely independent of drive amplitude or dissipation in the circuit. Since the dimension of the Poincaré section is 1.58, it can be inferred that the dimension of the actual flow in phase space is 2.58. The fractional value of the pointwise dimension is an indicator that the time-dependence of the circuit output is probably chaotic.

The Lyapunov exponents[4] are a better indicator of chaotic behavior. By definition, if an attractor has a positive Lyapunov exponent it is chaotic, that is, two points initially close together will diverge from each other exponentially with time, producing a sensitive dependence on initial conditions. The largest exponent is measured by finding the average rate at which nearby points diverge from each other. This immediately suggests a simple algorithm for computing the exponent from the data: choose a point $\{I_j, I_{j+1}, I_{j+2}\}$ at random, find its nearest neighbor and then follow their separation, $s_j(t)$, until it becomes a significant fraction of the "radius" of the attractor. (The last restriction is needed because the finite size of the attractor prevents the separation from increasing to arbitrarily large values.) The separation is then fit to the functional form,

$$s_j(t) \sim e^{\Lambda_j t}$$

for t not too large. The largest Lyapunov exponent, $\lambda_1$, is the average value of the $\Lambda_j$ of many random initial points on the attractor. The measured value of $\lambda_1$ was $0.33 \pm 0.01$, again with little dependence on either drive amplitude or dissipation. While this algorithm does not conform to the strict definition of the Lyapunov exponent, which requires that $t \rightarrow \infty$, it should give a good estimate of the exponent.

Direct observation of folding of the attractor is good visual evidence for the chaotic nature of an attractor. An examination of the attractor at a microscopic level which reveals more and more layers to the attractor as the magnification is increased would be clear evidence for chaos. However, this is not possible in an experiment since the precision is limited by the finite resolution of the equipment, in this case an 8 bit analog-to-digital converter. After only a few steps of magnification the layering that becomes apparent is due to the instrument and not the folding of the attractor. It is possible to observe large scale folding of the attractor even without high instrument resolution by following the development of the attractor as the sampling phase is varied from $0°$ to $360°$. The evolution of the attractor as the sampling phase is stepped by $45°$ increments can be clearly seen in Figure 3. Folding of the attractor becomes apparent if one follows the arm of the attractor labeled a. After $315°$ of phase shift it has clearly turned into arm b. Similarly, arm b transforms into arm c, and c into a (although in a much more obscure fashion). This final test, in conjunction with the measurements of a positive Lyapunov exponent and a fractional dimension, clearly establish the attractor of this simple circuit to be chaotic.

A full model[5] of the resistor, inductor, diode circuit results in a nonlinear second order differential equation for the charge stored in the diode of the form,

$$Q''(t) + f(Q)Q'(t) + g(Q) = A\sin(\omega t).$$

This model includes both the diode conductance and charge storage effects of the pn junction, which are different under forward and reverse bias. The nonlinear force term, $-g(Q)$, can be derived from the highly asymmetric effective potential shown in Figure 4. The restoring force is very stiff under reverse bias $(Q > 0)$, principly due to the conductance of the diode, and almost constant under forward bias of the junction $(Q < 0)$. The coefficient of friction, $f(Q)$, is a step function, but is constant over most of the range of Q. Numerical integration of this differential equation gives good quantitative agreement with the measured behavior of the circuit including

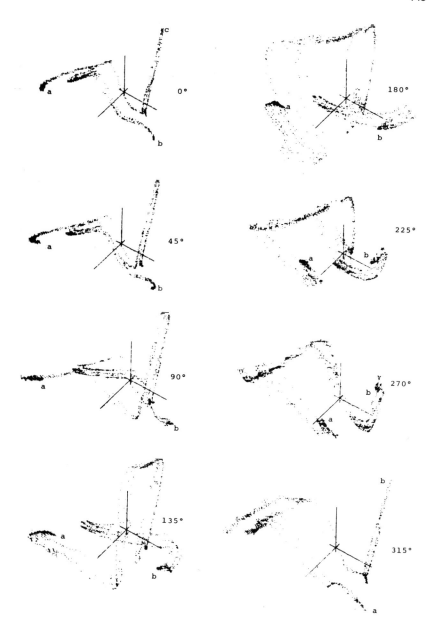

FIG. 3.    Folding of the chaotic attractor of Figure 1a at $V_d = 2870$ illustrated by changing the sampling phase in 45° steps.

the bifurcation diagram with hysteresis, the period-adding sequence, and the shape and evolution of the attractor with increasing drive amplitude. However, the calculations are time consuming and unedifying. This naturally leads one to simplify the equation of motion to arrive at one that is more tractable and understandable.

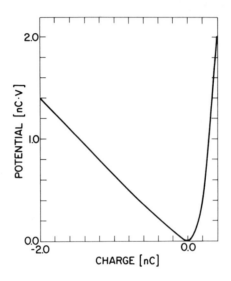

FIG. 4.    Effective potential of the nonlinear differential equation describing the diode circuit.

A simplified equation is,

$$\varsigma''(\phi) + \gamma\varsigma'(\phi) + 1 = \rho\sin(2\pi\phi), \quad \varsigma(\phi) \geq 0. \tag{1}$$

The original nonlinear velocity damping has been replaced with a uniform velocity damping and the nonlinear force term, $g(Q)$, replaced by a constant force normalized to 1. A nonlinearity similar to the stiff one-sided potential has been introduced to this second order linear differential equation by the boundary condition, $\varsigma(\phi) \geq 0$. (Other models and maps have also been proposed.[6]) The solution of this equation is elementary and given by,

$$\varsigma(\phi) = -\frac{\rho}{2\pi\sigma}\sin 2\pi(\phi + \theta) + \alpha e^{-\gamma\phi} - \frac{\phi}{\gamma} + \beta \ , \tag{2a}$$

$$\alpha = -\frac{1}{\gamma}\left[\varsigma_0' + \frac{1}{\gamma} + \frac{\rho}{\sigma}\cos 2\pi(\phi_0 + \theta)\right]e^{\gamma\phi_0} , \tag{2b}$$

$$\beta = \varsigma_0 + \frac{\phi_0}{\gamma} + \frac{\rho}{2\pi\sigma}\sin 2\pi(\phi_0 + \theta) - \alpha e^{-\gamma\phi_0} , \tag{2c}$$

$$\sigma = \sqrt{\gamma^2 + (2\pi)^2} , \tag{2d}$$

$$\theta = \frac{1}{2\pi}\tan^{-1}\frac{\gamma}{2\pi} , \tag{2e}$$

and the initial conditions $\varsigma_0$, $\varsigma_0'$, and the phase $\phi_0$ are defined by

$$\varsigma_0 = \varsigma(\phi_0) = 0, \quad \varsigma_0' = \varsigma(\phi_0) .$$

The equation can be thought of as describing a ball bouncing up and down on a table under the influence of a constant gravitational restoring force, velocity damping due to the air, and some (nongravitational) sinusoidal driving force, i.e., this is the equation of an impact oscillator. The ball bounces into the air, oscillating according to the driving force, and eventually falls back to the table under the influence of gravity. At impact ( $\varsigma(\phi) = 0$), the velocity reverses itself with some coefficient of restitution, and the ball continues on its way. If the coefficient of restitution is chosen to be zero so that the ball loses all its energy at impact, the resulting motion will be periodic with a period which depends on the driving parameter, $\rho$, but never chaotic. It is more interesting to make the coefficient of restitution equal to 1 so that the ball reverses its velocity on impact. This is also a better model of the diode circuit. A bifurcation diagram of the equation is shown in Figures 5a and 5b. Here $\varsigma(\phi)$ was sampled whenever $\phi = 0.1$ mod 1 was true. There is a clear resemblance between this and the bifurcation diagram of the data, including the period-adding sequence and the hysteresis at the lower window boundaries.

The period-adding sequence can be understood quite simply. As the driving force is increased, the mean impact velocity, and hence launch velocity of the ball, increases. This allows it to stay in the air for an ever larger number of oscillations of the driving force. For some value of $\rho$, the time in the air will exactly match $k$ cycles of the driving force, and a periodic window will appear. The velocity and

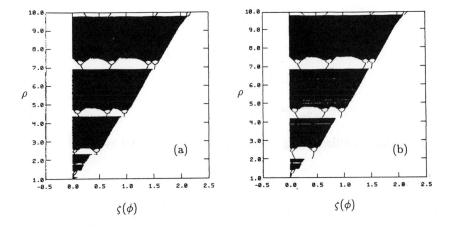

FIG. 5.     Bifurcation diagram of equation 1 plotting $\varsigma(\phi)$ versus $\rho$ whenever $\phi = 0.1$ mod 1. (a) $\rho$ increasing slowly with time, and (b) slowly decreasing with time. In both figures $\gamma = 1.0$ .

phase of the ball at impact can be analytically calculated for the large windows of period k, and are

$$\varsigma_k' = \frac{k}{2} \quad , \tag{3a}$$

$$\cos 2\pi (\phi_k + \theta) = \frac{\sigma}{\rho} \left( \frac{k}{2} \frac{1 + e^{-\gamma k}}{1 - e^{-\gamma k}} - \frac{1}{\gamma} \right) \quad . \tag{3b}$$

Since the cosine has a magnitude less than or equal to one, Equation 3b implies that there is a minimum value of $\rho$,

$$\rho_k = \sigma \left( \frac{k}{2} \frac{1 + e^{-\gamma k}}{1 - e^{-\gamma k}} - \frac{1}{\gamma} \right) \quad ,$$

for which the period k motion is stable. This is the simplest motion possible of period k for which there is exactly one impact for k oscillations of the driving force. More complicated periodic motions are possible which involve multiple impacts during one full k cycle. For example, there are two period 2 cycles, a single and a double impact cycle which are both observed in the experiment. In general, the number of possible period k motions equals the number of distinguishable ways of writing k as a sum of positive integers.

Since the solution of Equation 1 can be computed analytically between impacts given the initial conditions, it is possible to reduce the motion to a two dimensional map of the phase and velocity at impact,

$$\phi_{n+1} = \Phi(\phi_n, \varsigma'_n) \bmod 1 \ ,$$

$$\varsigma'_{n+1} = Z(\phi_n, \varsigma'_n) \ .$$

It is easy to compute the map for a given value of $\rho$ and $\gamma$ by using a root-finding routine to find successive zeroes of Equation 2. Inside a periodic window the two solutions to Equation 3 represent a stable and unstable fixed point. These points materialize at a saddle-node bifurcation when $\rho = \rho_k$, and both become unstable with increasing $\rho$ when an eigenvalue of the map's Jacobian crosses -1 and period doubling occurs. The computed transitions agree well with the boundaries in the bifurcation diagram of Figure 5b, made with $\rho$ decreasing from large values.

Comparison of the low edges of the periodic windows of figures 5a and 5b shows that boundary is different when the bifurcation diagram is made by increasing the value of $\rho$. This can be explained by referring to Figure 6, which is a map of $\phi_n$ and $\varsigma'_n$ inside the period 2 zone of hysteresis. The dotted bands in the upper part of the figure are the basin of attraction of the stable fixed point and the complicated wavy structure at the bottom of the figure is the chaotic attractor of the map. The hysteresis is due to the simultaneous coexistence of the fixed point and the attractor. For $\rho < \rho_2$ the fixed point does not exist, and the orbit of the motion is confined to the chaotic attractor. Even when $\rho$ becomes greater than $\rho_2$ this remains true because the attractor and the basin of the fixed point are well separated. Finally, when the control parameter has increased to a critical value, $\rho_c$, a crisis[7] occurs, and the chaotic attractor touches the basin of the fixed point, loses its stability, and all subsequent motion becomes periodic. If the value of $\rho$ is now decreased, the chaotic attractor detaches itself from the basin of the fixed point, but the orbit remains trapped at the fixed point and periodic until the saddle-node bifurcation at $\rho_2$ occurs. The abrupt spreading of the window at $\rho = 2.65$ after the period

148

doubling cascade has been completed is due to the occurence of a second crisis, the chaotic attractor expands because it collides with a three impact, period six unstable fixed point. The behavior of the boundaries of the other periodic windows can be explained in a similar manner.

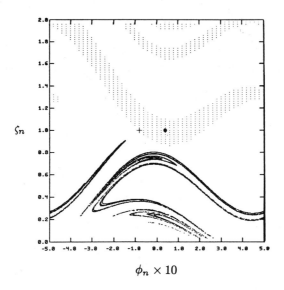

$\phi_n \times 10$

FIG. 6. Iterated map of the impact values of $\phi_n$ and $\varsigma_n'$, $\rho =$ 2.15, $\gamma = 1.0$ . The upper dotted bands are the basin of attraction of the stable period 2 fixed point (circle). The wavelike figure at the bottom is the coexisting chaotic attractor and the unstable period 2 fixed point is marked by a cross.

I would like to thank the Office of Naval Research, the Alfred P. Sloan Fund for Basic Research, and the National Science Foundation for support.

# References

1)    P. S. Linsay, Phys. Rev. Lett 47, 1349(1981); J. Testa, J. Perez, and C. Jeffries, Phys. Rev. Lett. 48, 714(1982); C. Jeffries and J. Perez, Phys. Rev A26,2117(1982); H. Ikezki, J. S. DeGrassie and T. H. Jensen, Phys. Rev A28, 1201(1983); J. Cascais, R. Dilao, and A. Noronha Da Costa, Phys. Lett. 93A, 213(1983); T. Klinker, W. Meyer-Ilse, and W. Lauterborn, Phys. Lett. 101A, 371(1984); P. S. Linsay, Phys. Lett. 108A, 431(1985).

2)    M.J. Feigenbaum, J. Stat. Physics 19, 25(1978).

3)    J.D. Farmer, E. Ott, and J. A. Yorke, Physica 7D, 153(1983); Y. Termonia and Z. Alexandrowicz, Phys. Rev. Lett. 51, 1265(1983); J. Guckenheimer and G. Buznya, Phys. Rev. Lett. 51, 1438(1983); A. Brandstater, et al., Phys. Rev. Lett. 51, 1442(1983)

4)    E. Ott, Rev. Mod. Phys. 53, 655(1981); J. D. Farmer, Physica 4D, 366(1982); A. J. Lichtenberg and M.A. Lieberman, Regular and Stochastic Motion, Springer-Verlag, New York, 1983.

5)    S. D. Brorson, D. Dewey, and P. S. Linsay, Phys. Rev. A28, 1201(1983).

6)    R. W. Rollins and E. R. Hunt, Phys. Rev. Lett. 49, 1295(1982); A. Azzouz, R. Duhr, and M. Hasler, IEEE Trans. CAS-30, 913(1983); R. Van Buskirk, LBL-17868, 1984; E. R. Hunt and R. W. Rollins, Phys. Rev. A29, 1000(1984); T. Matsumoto, L. O. Chua, and S. Tanaka, Phys. Rev. A30, 1155(1984); T. Yoon, J. Song, S. Shin, and J. Ra, Phys. Rev. A30, 3347 (1984).

7)    C. Grebogi, E. Ott, and J. A. Yorke, Phys. Rev. Lett. 48, 1507(1982), and Physica 7D, 181(1983)

# PHASE SPACE STRUCTURE AND DIFFUSION

Allan J. Lichtenberg

Department of Electrical Engineering and Computer Sciences
and the Electronics Research Laboratory
University of California, Berkeley, CA  94720

## ABSTRACT

The effect of resonant behavior on phase space is
described.  The significance of stochastic layers
in enhancing diffusion is discussed.  Diffusion
along resonance layers that arises in more than
two degrees of freedom, e.g., Arnold diffusion
and modulational diffusion, is described, and
illustrated with an example.

## 1.    Introduction

It is well known that Hamiltonian systems with one degree of
freedom $H(p,q)$ are integrable.  For two degrees of freedom $H(p_1,p_2,q_1,$
$q_2)$ integrability is exceptional.  In general, resonances between the
two degrees of freedom lead to the formation of resonance layers in the
action space.  Within each resonance layer, chaotic motion appears.
Energy conservation prevents large excursions of the motion along the
layer.  Only motion across the layer is important.  For an integrable
system with a weak perturbation the chaotic layers are isolated by
Kolmogorov-Arnol'd-Moser (KAM) surfaces.  Thus motion from one layer
to another is forbidden.  For strong perturbations, resonance layers
can overlap, the intervening KAM surfaces being destroyed.  A more
global chaotic motion then develops, leading to large excursions in
both actions over long times.

For three or more degrees of freedom, strong perturbations also
lead to overlap of resonance layers and globally chaotic motion.  However, for weak perturbations, two new effects appear:

1.    Resonance layers are no longer isolated by KAM surfaces.

Generally, the layers intersect, forming a connected web dense in the action space.

2. Conservation of energy no longer prevents large chaotic motions of the actions along the layers over long times. As a result, large, long-time excursions of the actions along resonance layers are generic in systems with three or more degrees of freedom. The interconnection of the dense set of layers ensures that the chaotic motion, stepping from layer to layer, can carry the system arbitrarily close to any region of the phase space consistent with energy conservation.

The basic technique for analyzing these phenomena starts with the isolation of a single resonance. A transformation is made to action-angle variables for which one angle is slowly varying near the reso-- nance. The method of averaging is then employed to average over the fast variables, reducing the problem to a single degree of freedom for which the solution is obtainable. By reintroducing the higher order terms the process can be repeated to isolate a $2^{nd}$ order resonance between the primary motion and the slow oscillation about the resonance. Since the resonances are derived from Fourier expansions, all resonances have the same form, when expressed in appropriate variables. The process may be called resonance **renormalization.**

We consider first the technique of resonance renormalization and overlap. We then consider the effect of resonances on diffusion of the action space. Finally, we discuss the phenomenon of diffusion along resonance layers (Arnold diffusion) in systems with more than two degrees of freedom, again illustrating the behavior with an example. The treatment in the following sections will be quite brief, with references given for the longer calculations. For simplicity, most references will refer to the monograph <u>Regular and Stochastic Motion</u>,[1] rather than to the original sources which are referenced there. Topics finished subsequently to the monograph are referenced separately.

2.    Resonance Renormalization and Resonance Overlap

We use as our initial mapping the Chirikov-Taylor map or "standard map"[1] (Ref 1, Sec. 4.1b)

$$I_{n+1} = I_n + K \sin \theta_n \; , \qquad (1)$$

$$\theta_{n+1} = \theta_n + I_{n+1} \; . \qquad (2)$$

Other mappings can often be locally represented by the standard map. For example a mapping which approximates many physical systems, the Fermi acceleration map

$$u_{n+1} = u_n + \sin \theta_n \; , \qquad (3a)$$

$$\theta_{n+1} = \theta_n + \frac{2\pi M}{u_{n+1}} \qquad (3b)$$

approximates the standard mapping in the neighborhood of values of action

$$\frac{M}{u_0} = m, \; m \text{ integer.} \qquad (4)$$

Expanding around $u_0$,

$$u_{n+1} = u_0 - \frac{M}{u_0^2} \Delta u_{n+1} , \qquad (5)$$

**the substitutions**

$$I_n = - \frac{2\pi M}{u_0^2} \Delta u_n \; , \qquad (6)$$

$$K_n = - \frac{2\pi M}{u_0^2} \qquad (7)$$

reproduce the standard map.

The fixed points for a single iteration of (1) and (2) are given by

$$I_0 = 2\pi m \; , \qquad (8a)$$

$$\sin \theta_0 = 0, \qquad (8b)$$

and expanding about this fixed point the eigenvalues of the linearized map are obtained:

$$\cos \sigma = \frac{1}{2}(2 - K \cos \theta_0) , \qquad (9)$$

where $\sigma$ is the phase shift per mapping period. The motion is linearly stable for

$$|\cos \sigma| < 1 \qquad (10a)$$

i.e., for

$$-4 < K \cos \theta_0 < 0. \qquad (10b)$$

We can make the mapping into a flow by introducing the periodic $\delta$-function, with Hamiltonian representation[1] (Ref. 1, Sec. 3.4e),

$$H = \frac{I^2}{2} - K \cos \theta \sum_m e^{i2\pi m\tau} , \qquad (11)$$

where $\tau$ is a normalized time that advances by unity for each mapping iteration. The summation is the Fourier representation of a $\delta$-function For $d\theta/d\tau \ll 2\pi$, an average over $\tau$ leaves only the $m = 0$ term

$$H = \frac{I^2}{2} - K \cos \theta = H_0. \qquad (12)$$

This is just the pendulum Hamiltonian.

$$H = G \frac{I^2}{2} - F \cos \theta = H_0$$

for the special case $F = K$, $G = 1$. The amplitude in action, of the phase space "island" associated with this motion is

$$\Delta I_m = 2(F/G)^{1/2} .$$

The pendulum motion can also be transformed to action-angle form

$$\bar{H} = \bar{H}_0(J). \qquad (13)$$

For particular values of $J$ there are resonances in (11):

$$p\omega + 2\pi m = 0. \qquad (14)$$

For these resonances the phase flow given by (13) is not correct. We reintroduce the summation in (11) but express the Hamiltonian in terms of the action-angle variables $J$, $\theta$ as

$$H = H_0(J) + K \sum_{\substack{m(m \neq 0)}} \sum_{\ell} V_\ell(J) e^{i(\ell\phi + 2\pi m\tau)} , \tag{15}$$

where we have expanded $\cos \theta$ in its Fourier coefficients $V_\ell$.

The renormalization procedure is to obtain a Hamiltonian valid, locally, near a next higher order resonance where the value of action $J_p$ satisfies (14). We write an approximate Hamiltonian $\Delta H$ in terms of action $\Delta J = J - J_p$:

$$\Delta H = \frac{\partial H_0}{\partial J} \Delta J + \frac{\partial^2 H_0}{\partial J^2} \frac{\Delta J^2}{2} + K \sum_{\substack{m \\ (m \neq 0)}} \sum_{\ell} V_\ell(J_p) e^{i(\ell\phi + 2\pi m\tau)} , \tag{16}$$

with the partial derivatives and $V_\ell$ evaluated at $J_p$. The first partial will be transformed away by virtue of evaluating at the resonance. We restrict our attention to the integer resonances, $m = 1$, since for higher values of $m$ the Fourier amplitudes are small. A canonical transformation

$$\hat{\phi} = p\phi + 2\pi\tau ,$$

$$\hat{J} = J/p , \tag{17}$$

$$\hat{H} = H + 2\pi\hat{J}$$

produces a Hamiltonian with slowly varying phase near resonance, i.e., $\dot{\hat{\phi}} = 0$ at resonance, and averaging over the fast phase we obtain the local Hamiltonian

$$\Delta\hat{H} = p^2 \frac{\partial^2 H_0}{\partial J^2} \frac{\Delta J^2}{2} + 2KV_p(J_p)\cos \hat{\phi} , \tag{18}$$

which can be written in the form

$$H_2 = G_S \frac{\bar{J}^2}{2} + 2F_S \cos \bar{\phi} , \tag{19}$$

with $\bar{J} = \Delta\hat{J}$, $\bar{\phi} = \hat{\phi}$, and

$$F_S = KV_p \qquad G_S = p^2 \frac{\partial^2 H_0}{\partial J^2} . \tag{20}$$

Equations (19) and (20) give the amplitude of a resonant second order island chain,

$$\Delta \hat{I}_M = 2(F_s/G_s)^{1/2} \tag{21}$$

completing the renormalization   (see Ref. 2 for a more complete calculation).

In second order, we have the universal relations for island 'overlap' i.e., no KAM surfaces between the $\ell$ and $\ell + 1$ islands   (Ref 1, Chapter 4).

$$\frac{\Delta I_{M\ell} + \Delta I_{M\ell+1}}{|I_{0\ell} - I_{0\ell+1}|} \geq \frac{2}{3} \; . \tag{22}$$

For nearly equal size islands, substituting from the above, we obtain the condition for stochasticity joining two island chains:

$$36 F_s G_s / FG \geq 1 \; . \tag{23}$$

The overlap of $2^{nd}$ order resonances determines the thickness of the stochastic layer that exists generically around the separatrices of each resonance.

In Fig. 1 we illustrate the standard mapping with its main island and surrounding stochastic layer associated with the overlap of $2^{nd}$ order islands.  For the value of K = 1.19 chosen for this computation the 4-iteration island in rotation (shown in the box) is self-similar with the main island chain, and also satisfies (23).  Thus we expect that the last KAM surface between the main islands will also be destroyed, leading to diffusion across the entire mapping.  This is indeed the case, and in fact the value of K satisfying (23) is $K \geq 1$.

3.   Diffusion Across a Resonance Layer

The diffusion rate across a stochastic layer can be calculated by first modifying the phase space so that it looks locally like a standard mapping   (see Ref. 1, Sections 3.5 and 4.1) and then using the long-time diffusion calculation in the presence of correlations and **islands**   (see Ref. 1, Sections 5.4d and 5.5).  We illustrate this procedure with the local transformation from the Fermi map to the standard

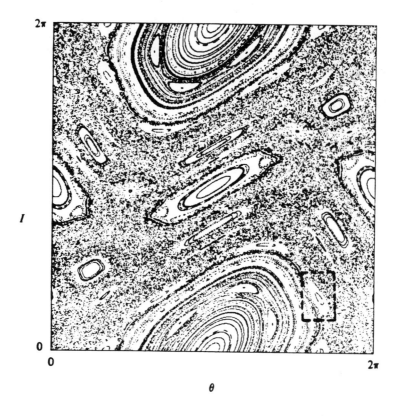

Fig. 1. The standard map for K = 1.9. In the dashed box is a second order island whose renormalized standard mapping has K = 1.9.

map as given by Eqs. (6) and (7). This correspondence is shown schematically in Fig. 2. The Fermi map in the neighborhood of $u_b$, the
velocity at which the first KAM curve spans the space, corresponds to
the standard map at $K \approx 1$. The separatrix layer has a mapping representation similar to the Fermi map, described more completely in Ref. 1,
Sec. 3.5b. To determine the diffusion rate across such a layer or
across a mapping such as the Fermi mapping, the procedure is to compute the long time diffusion coefficient for the standard map as a
function of the parameter K, and then use this velocity-dependent diffusion coefficient in a Fokker-Planck equation to calculate the diffusion. An example of the result of this procedure is shown in Fig. 3,
comparing the variance of the time-dependent distribution obtained
from the Fokker-Planck solution, with the variance obtained from the
iterated Fermi map for a large number of initial conditions, at each
velocity u. A more complete description of the method and results is
given in Ref. 3.

For regions of the phase space in which a local approximation to
the standard map gives $K < 1$, either within a separatrix layer or for
a macroscopic mapping such as the Fermi map, KAM curves prevent continuous intrinsic diffusion. However, if there is extrinsic noise in
the system then diffusion exists over the entire phase plane due to
this noise. This can be seen by adding a random jump $\xi$ to the phase
variable in the standard map, such that (2) becomes

$$\theta_{n+1} = \theta_n + I_{n+1} + \xi_n , \tag{24}$$

where $\xi_n$ has a Gaussian probability distribution with mean square
deviation $\sigma$. On an isolating KAM curve a jump in $\theta$ due to $\xi$ leads to
a jump in the action

$$\delta I = |I_m \cos \theta| \xi , \tag{25}$$

such that when averaged over the Gaussian distribution gives a spread
in action

$$(\delta I)^2_{rms} = \frac{1}{2} I_m^2 \sigma. \tag{26}$$

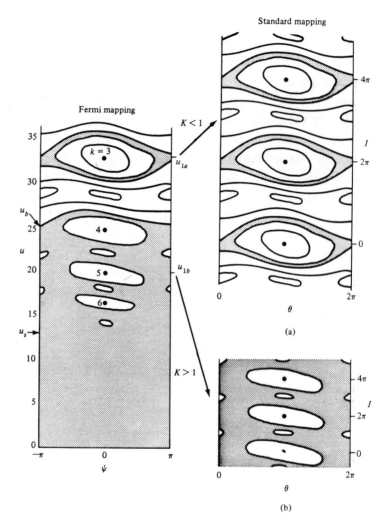

Fig. 2.  Local approximation of the Fermi mapping by the standard mapping.  (a) Linearization about $u_{1a}$ leading to K small and local stochasticity; (b) Linearization about $u_{1b}$ leading to K large and global stochasticity.

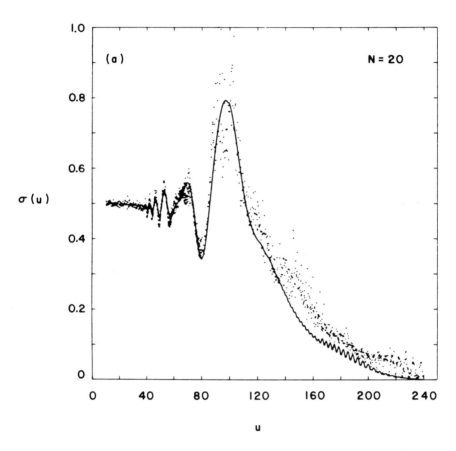

Fig. 3. The variance as a function of initial velocity (action) after 20 mapping iterations. Solid line is from integration of the Fokker-Planck equation. The quasilinear diffusion (no phase correlation) gives $\sigma(u) = 0.5$. The first KAM barrier $u = u_b$ is at $u_b = 250$. (M = 10,000 in Eq. (3).)

From the averaged Hamiltonian, as in (12), but taking the rotation rather than the libration orbits, a typical amplitude is $I_m \approx K/2$, giving a diffusion

$$D \approx \frac{(\delta I)^2_{rms}}{1} = \frac{K^2}{8} \sigma, \tag{27}$$

If there are islands of significant size, surrounded by stochastic layers, then the diffusion across these layers, typically occurs on a time scale much more rapid than that of the extrinsic diffusion. We illustrate this effect in Fig. 4. Considering a phase point that scatters into the stochastic region of the main island, shown as point A, then the subsequent trajectory shown by arrows $A \rightarrow B(B') \rightarrow C'$ takes the phase point around the island to $C'$ where it may again be scattered onto an isolating KAM curve. If the time for oscillation T, is small compared to $1/\sigma$ [which would be generally satisfied for small $\sigma$ and $K = O(1)$, since $T = O(K^{-1/2})$], then the phase space in which diffusion occurs is shrunk by some factor proportional to the island width. Note that the diffusion from the trajectory $A \rightarrow C'$ can be either outward as described above, or inward onto the libration orbits surrounding the main island. These latter orbits do not contribute to a fast diffusion. Thus there are two classes of particles present in the system that contribute to the overall diffusion rate: a fast class that streams across the main island and a slow class that diffuses into the island. We take as a simple estimate that the diffusion rate of particles diffusing fast to those diffusing slow is inversely proportional to the square of the action distance traversed:

$$\frac{D_f}{D_s} = \frac{(2\pi)^2}{[2\pi - 2K^{1/2}]^2}, \tag{28}$$

where $2K^{1/2}$ is the width of the large island in Fig. 4, and we ignore the effects of other islands. We further consider that half the trajectories incident on A are trapped in the half-island oscillation from A to C, and that half of these are detrapped at C. Thus one-fourth of the trajectories are in the fast class, streaming across

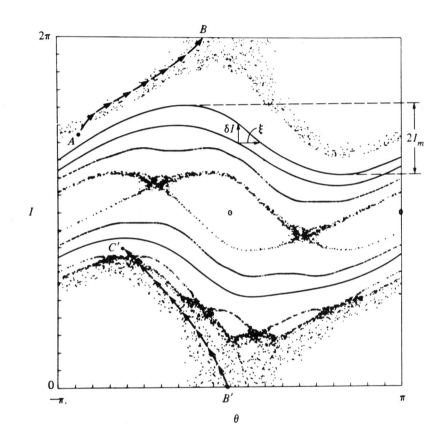

Fig. 4.    Phase plane for the standard mapping, illustrating a streaming trajectory that produces enhanced extrinsic diffusion
for K ≲ 1.   The θ-coordinate is shifted by π from Fig. 1.

from A to C', such that the effective diffusion is

$$D_{eff} = \frac{1}{4} D_f + \frac{3}{4} D_s. \tag{29}$$

This heuristic calculation gives reasonable agreement with more exact calculations (see Ref. 1, Sec. 5.5b) even for K someqhat greater than 1, where the intrinsic stochasticity is global and would proceed for $\sigma = 0$. Near KAM surfaces that existed for K < 1, the trajectories still tend to follow these just-destroyed surfaces over long times. Thus the overall diffusion rate is still governed by the value of $\sigma$ across these regions of the phase plane.

4.    Arnold Diffusion

We consider the geometry of stochastic layers in the 2N-dimensional phase space. The layers are surfaces having dimension 2N-1. The KAM surfaces, being perturbed tori, are N-dimensional. The interconnection of resonance layers into the Arnol'd web can then be understood geometrically. For $N \geq 3$, the (2N-1)-dimensional resonance surfaces cannot be isolated from one another by N-dimensional KAM surfaces. The "Arnold" diffusion along the separatrix layers of the interconnected resonances is generic to systems with more than two degrees of freedom.

A simple example of a system illustrating Arnol'd diffusion is that of a ball bouncing back and forth between a smooth wall at z = h and a fixed wall that is rippled in two dimensions, x and y, at z = 0. The surface of section is given in terms of the ball positions $x_n$ and $y_n$ and the trajectory angles $\alpha_n = \tan^{-1} v_x/v_z$ and $\beta_n = \tan^{-1} v_y/v_z$, just before the $n^{th}$ collision with the rippled wall. The ball motion is shown schematically, and variables in the x, z-plane are defined in Fig. 5. Assuming that the ripple is small, the rippled wall may be replaced by a flat wall at z = 0 whose normal vector is a function of x and y, analogous to the idea of a Fresnel mirror. The simplified difference equations exhibit the general features of the exact equations and may be written in explicit form

$$\alpha_{n+1} = \alpha_n - 2a_x k_x \sin k_x x + \mu k_x \gamma_c \ ,$$

$$x_{n+1} = x_n + 2h \tan \alpha_{n+1} \ ,$$

$$\beta_{n+1} = \beta_n - 2a_y k_y \sin k_y y + \mu k_y \gamma_c \ ,$$ (30)

$$y_{n+1} = y_n + 2h \tan \beta_{n+1} \ ,$$

where $\gamma_c = \sin(k_x x + k_y y)$, $a_x$, and $a_y$ are the amplitudes of the ripple in the x- and y-directions, respectively, and $\mu$ is the amplitude of the diagonal ripple and represents the coupling between the x- and y-motions.

A typical numerical calculation showing Arnol'd diffusion in the coupled system is given in Fig. 6. The surface of section for the system is four-dimensional $(\alpha,x,\beta,y)$, which we represent in the form of a pair of two-dimensional plots $(\alpha,x)$ and $(\beta,y)$. Thus, two points, one in $(\alpha,x)$ and one in $(\beta,y)$, are required to specify a point in the four-dimensional section. In Fig. 3 the two plots are superimposed for convenience, and x and y have been normalized to their respective wavelengths, $2\pi/k_x$ and $2\pi/k_y$. The initial condition (Fig. 6a) has been chosen on an island encircling the central resonance in x, and within the thin separatrix layer for y. This corresponds to an initial adiabatic motion in x, well confined in the valley, while in y the motion just reaches or passes over a hill. We observe numerically that the y-motion is confined to its separatrix layer until the x-motion reaches its own separatrix layer. The successive stages of the diffusion of the $\alpha$-x motion are shown in Fig. 6b, c, and d, respectively. In the absence of coupling ($\mu=0$), the motion in the $\alpha$-x plane is confined to a smooth closed curve encircling the central resonance. For a finite coupling, $\alpha$ and x diffuse slowly because of the small randomizing influence of the stochastic $\beta$-y motion. The $\alpha$-x diffusion is motion along the $\beta$-y stochastic layer; that is, it is the Arnol'd diffusion. The diffusion is shown for $1.5 \times 10^5$, $3.5 \times 10^6$, and $10^7$ iterations of the mapping. At this time the $\alpha$-x motion has diffused out to its own thin separatrix layer. Continued iteration of the

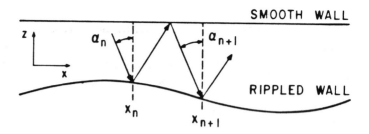

Fig. 5.  Motion in two degrees of freedom, illustrating the definition of the trajectory angle $\alpha_n$ and bounce **position** $x_n$ just
before the $n^{th}$ collision with the rippled wall.

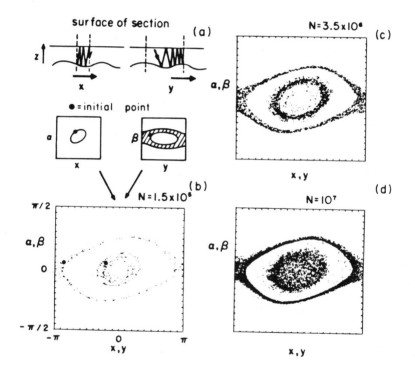

Fig. 6.  Thin layer diffusion.  Initial conditions are close to the
central resonance in the $\alpha$-$x$ space and within the separatrix
stochastic layer in the $\beta$-$y$ space.  Parameters are $\mu/h$
= 0.004; $\lambda_x : h : a_x$, and $\lambda_y : h : a_y$ as 100 : 10 : 2.

mapping shows that the trajectory point diffuses over the entire $\alpha$-x plane. In particular, the change of direction from diffusion along the $\beta$-y separatrix layer to diffusion along the $\alpha$-x separatrix layer has been observed numerically.

The rate of diffusion for this example has been calculated, using the stochastic pump model, and compared with the numerical results, giving good agreement (see Ref. 1, Section 6.2). The diffusion rates can be very sensitive, however, to resonance positions and thickness, so that accurate diffusion calculations are difficult in most practical cases.

There is an interesting, related case in which the frequency of one of the two interacting degrees of freedom is slowly modulated in time. For a single modulated degree of freedom the slow variation is adiabatic, and KAM surfaces exist. With the addition of the second degree of freedom a multiplet layer is formed which can overlap, leading to diffusion across the layer and enhanced diffusion along the layer, known as modulational diffusion. The overlap requires sufficiently slow modulation, which is in contra-distinction to the normal adiabatic problem. The diffusion along a multiplet layer can be calculated, and is often strong compared to the diffusion along the separatrix layer of a single resonance (see Ref. 1, Section 6.2d).

If extrinsic noise is added to an autonomous system with more than two degrees of freedom, or a mapping (non autonomous system) with more than one degree of freedom, then the noise can drive the diffusion along the Arnold web, strongly enhanced by the resonances. The KAM curves of the intrinsic system no longer act as barriers in the manner that they did in Sec. 3 of this paper. This situation is considered and calculations of the resultant diffusion are made in Ref. 1, Sec. 6.3.

Acknowledgement

Research reported on in this paper includes joint work with M. A. Lieberman, N. W. Murray, and J. L. Tennyson. The work was sponsored in part by the Office of Naval Research Contract N00014-84-K-0367.

References

1) A. J. Lichtenberg and M. A. Lieberman, <u>Regular and Stochastic Motion</u>, Springer-Verlag, New York, 1983.
2) A. J. Lichtenberg, Physica <u>14D</u> 387 (1985).
3) N. W. Murray, M. A. Lieberman, and A. J. Lichtenberg, "Corrections to Quasilinear Diffusion in Area Preserving Maps," Phys. Rev. A, to be published (1985).

A Low Dimensional Iterated Map with an

Apparent Continuum of Attractors

Henry D. I. Abarbanel,[1]

S. B. Buchsbaum,[2]

and J. D. Keeler,[3]

1.  Scripps Institution of Oceanography and Project for Nonlinear  Science;
UC  San  Diego;  La Jolla, CA  92093.  Work supported by U.S. Department of
Energy, Office of Basic  Energy  Sciences,  under  contract  DOE  DE  ASO3-
84ER13165  and  by the Office of Naval Research, Code 422PO, under contract
N00014-79-0472.

2.  Department of Physics; UC San Diego; La Jolla, CA  92093

3.  Department of Physics and Project for Nonlinear Science; UC San  Diego,
La Jolla, CA  92093.

ABSTRACT


In the investigation of two coupled area preserving maps of the unit square, we have found a four dimensional map which does not conserve phase space volume globally and which appears to have a continuum of attractors. The attractors range from quasiperiodic to chaotic as the map parameters are varied. In this report we present the map, some qualitative discussion of its features, a variety of pictures displaying the attractors, and some possible implications of a very large, possibly infinite number of basins of attraction. We also consider the possibility of basins of attraction of lower dimension than the phase space.

. .

# I. Introduction

The asymptotic behavior of maps not preserving phase space volume is characterized by attractors of enormous variety and structure.[1-5] The precise fashion in which an orbit visits the attracting set is sensitive to initial conditions, so one usually speaks of properties of the attracting set not dependent on such details. These properties include the dimension of the attractor,[6,7] variously defined,[8,9] and the asymptotic distribution of points on the attractor.[10] The systems which are familiar in the contemporary literature on strange attractors possess at most a few basins of attraction. The boundaries of these basins can be very rich and complicated, indeed, having themselves a fractal nature.[11,12] There appears to be no a priori reason to expect that the number of attractors even in a low dimensional dynamical system need be small. Indeed, any experiment where the detailed results seem to be irreproducible might very well have a large number of attractors with complicated basin boundaries separating them. This would imply that a slight change in the initial conditions would lead to different final states leading to a different set of observations each time the experiment is performed.

In this short note we will present a four dimensional iterated map living on the four torus which appears to have, for fixed parameters of the map, a different attractor for an infinite number of initial conditions. Some of these attractors are shown in Figures 1-3. We have only a few analytic or quantitative remarks to make about the map and do not understand the phenomena we are observing in any fundamental way. Nevertheless, we have performed numerous tests on the numerical reliability of the phenomena we are seeing and are convinced that this represents a heretofore uninvestigated interesting and novel aspect of dynamical systems. The system we present was created as a model of features of some physical systems, however, in itself it does not represent the behavior of any physical system known to us. To that extent it may represent a mathematical curiosity. If one can find a physical system in which this phenomenon does occur, an additional, perhaps undesired, form of unpredictability will have to be contended with in understanding the long time behavior of dynamical systems. In particular, if, after transients have died out, the final state of a dynamical system depends sensitively on initial conditions for a set of initial conditions which occupies a finite volume in phase space, the ability to predict the detailed outcome of an experiment is in much doubt. Even the conventional smoothed or averaged quantities not sensitive to specific orbits may need reexamination as meaningful indications of the properties of a system.[13]

With a system whose asymptotic state is sensitive to initial conditions we face another partially disappointing matter: the calculation of the final state is likely to be dependent on both the precise algorithm one uses to iterate the map as well as dependent on the machine one employs for the calculation. In other words, the attractors are not robust under small changes in the precise manner used for the numerical calculation. Even roundoff errors which can make an orbit visit the attractor in an isolated basin of attraction in a different fashion but not change the dimension of the observed attractor will here bump orbits onto completely different attractors.

This kind of dependence on numerical methods and perhaps even the machine on which one evolves the dynamical system may present as much of an opportunity for cryptography[14,15] and models of intelligence[16-20] as it presents a problem for the interpretation of laboratory experiments.[21] One may imagine the map itself to be publicly available. The message to be encoded and transmitted would be contained in properties of the attractor itself. The initial conditions would constitute the key for the message. The key would decode the message, that is reveal the attractor on which one might wish to perform some subsequent analysis, when used with the agreed upon algorithm for implementing the iterated map and when implemented on the same type of machine used by the sender.

Similarly, though even more speculatively, one may imagine a model of intelligence in which the information to be transmitted lies in the properties of the attractor which are reached by a built in rule (the map to be iterated) and that the message is carried in a compact and efficient form by a few initial conditions. On receiving these initial conditions, the receiver runs the map and reads the message. The mechanism is much the same as in the previous example, but the desire to hide the contents of the message would be absent.

We do not dwell further on the possible implications of the phenomenon we have uncovered. It is interesting and novel in its own right. The purpose of this note is to report on this phenomenon rather than to explain and exploit it. The majority of the paper will therefore be descriptive. In the next section we describe the map and give the motivation that led us to it. Following that we describe what we have observed and present many pictures to illustrate it. We conclude with some discussion of our results.

## II. The Iterated Map

Our motivation for studying the map we discuss here came from investigating two coupled volume preserving systems with rather different spatial and temporal scales. The origin of this problem lies in the attempt to understand the rates of transfer of physical quantities such as action or momentum from the slower, larger scale system to the faster. Transfer of energy from mesoscale oceanic flows to internal waves falls in this category. We have been so involved in the investigation of the curious phenomenon we have uncovered that we have not had the opportunity to make the interesting connection with this kind of physical problem, but we will return to it in another publication.

We chose to represent each of the two systems to be coupled by a two dimensional area preserving iterated map, called the standard map or the Chrikov-Taylor map.[22,23] This was done because previous studies of this map have been extensive[24,25] and have revealed much about its transition from regular to globally chaotic behavior. The two phase space coordinates of the map are called $p(n)$ and $q(n)$, where n is a label indicating the iteration or "time." They represent, for example, the momentum and angular coordinate of a pendulum perturbed at unit intervals by a delta function force. The generating function for the canonical transformation which takes one from $\{p(n), q(n)\}$ to $\{p(n+1), q(n+1)\}$ is

$$S(q(n), p(n+1)) = q(n)p(n+1) + \frac{1}{2}(p(n+1))^2 + \frac{k}{4\pi^2}\cos\,[2\pi q(n)] \cdot \tag{1}$$

The map arises from the rules[26]

$$p(n) = \frac{\partial S}{\partial q(n)}\,, \tag{2}$$

$$q(n+1) = \frac{\partial S}{\partial p(n+1)}\,, \tag{3}$$

and is

$$p(n+1) = p(n) + \frac{k}{2\pi}\,\sin\,[2\pi q(n)], \tag{4}$$

$$q(n+1) = q(n) + p(n+1)\,. \tag{5}$$

Since this map comes from a generating function, the volume in phase space is preserved at each iteration, as is easily checked. Furthermore, although the map is originally defined on $R^2$, because of the symmetries it possesses, one may consider it to lie on the unit two torus: $0 < q < 1$ and $0 < p < 1$ with no loss of information. The area is, of course, preserved locally by this restriction since the Jacobian of the transformation (1) remains unity. Because of the periodic nature of the nonlinear term, the map is invertible on the torus and thus preserves area on the torus globally. Various past studies of this map have shown that for the map parameter[27] $k = .9716$, the last isolating KAM surface disappears and there are orbits free to wander over the entire phase space; that is, the unit torus.

Now we wish to couple together[28-31] two of these maps with differing fundamental frequencies, by which we mean the rate at which q, the angle like variable increases at each iteration. Our map will involve two q's and two p's which we will call $q_s$, $p_s$, $q_b$, and $p_b$. The first system (called system) is the representation of the slower part of the coupled system; the other (the bath) is the faster. The transformation between $q_s(n)$, $p_s(n)$, $q_b(n)$, and $p_b(n)$ and their values at the next iteration labeled by n+1 arises from the canonical generating function $S\{q_s(n)$, $q_b(n)$, $p_s(n+1)$, $p_b(n+1)\}$ taken to be

$$S\{q_s(n), q_b(n), p_s(n+1), p_b(n+1)\} = q_s(n)p_s(n+1) + q_b(n)p_b(n+1)$$

$$+ \frac{1}{2} \omega_s(p_s(n+1))^2 + \frac{1}{2} \omega_b(p_b(n+1))^2 + \frac{k_s}{4\pi^2\omega_s} \cos [\ 2\pi q_s(n)\ ]$$

$$+ \frac{k_b}{4\pi^2\omega_b} \cos [\ 2\pi q_b(n)\ ] + bp_s(n+1)g[q_s(n) + q_b(n)] \ . \tag{6}$$

The coupling $bp_s(n+1)g[q_s(n) + q_b(n)]$ is chosen to make the resulting map rather simple. The function g, given below, was chosen to emphasize the tendency of phase space volumes to contract on the slower time scale. Once again this motivation may have little to do with the implications of the final map.

The rule above for generating the map then yields

$$p_s(n+1) = \frac{p_s(n) + \dfrac{k_s}{2\pi\omega_s} \sin [\ 2\pi q_s(n)\ ]}{1 + bg'[q_s(n) + q_s(n)]} \ , \tag{7}$$

$$q_s(n+1) = q_s(n) + \omega_s p_s(n+1) + bg[\ q_s(n) + q\ (n)\ ]\ , \tag{8}$$

$$p_b(n+1) = p_b(n) + \frac{k_b}{2\pi\omega_b}\ \sin\ [2\pi q_b(n)]\ -\ bp_s(n+1)g'(q_b(n) + q_s(n))\ , \tag{9}$$

$$q_b(n+1) = q_b(n) + \omega_b p_b(n+1)\ , \tag{10}$$

where $g'(x) = dg(x)/dx$.

The map as written is from $R^4$ to itself and preserves volume globally for smooth g. If the coupling parameter b were zero, we would naturally take the variables to all lie in the interval 0 to 1 and thus consider the whole map on the torus $T^4$. If we do that here, as we, of course, may choose to do, we are not guaranteed that the resulting map will globally preserve volume, though locally the Jacobian of the transformation (7-10) is always unity. Indeed, we must have the function g periodic in its arguments as well as smooth enough to guarantee that the map is not only smooth but also invertible. If those conditions were met, the map would be globally volume preserving, even on $T^4$.

Nevertheless, we chose the function g to be

$$g(x) = \frac{1}{2}\ \left[\ x + \sin\ [\ 4\pi x\ ]\ /\ 4\pi\ \right]\ , \tag{11}$$

which does not lead to global volume preservation, though it did satisfy other requirements of no special importance here. Further we restrict the phase space by considering each of the operations taking us from coordinates at "time" n to "time" n+1 to be modulo unity. Having taken this apparently innocent step, we are led to the remarkable results we have promised earlier.

### III.  The Observed Phenomena

This map, once g(x) is given, has only four independent parameters. We may exhibit this by defining the frequency ratio

$$f = \omega_b/\omega_s \tag{12}$$

and scaling each of the p(n)'s by

$$\omega_b p_b(n) \rightarrow p_b(n) , \qquad\qquad \omega_s p_s(n) \rightarrow p_s(n) . \tag{13}$$

This results in the iterated map

$$p_s(n+1) = \left\{ \frac{p_s(n) + \frac{k_s}{2\pi} \sin [\ 2\pi q_s(n)\ ]}{1 + bg'[\ q_s(n) + q_b(n)\ ]} \right\} \bmod 1 , \tag{14}$$

$$q_s(n+1) = \{q_s(n) + p_s(n+1) + bg[\ q_s(n) + q_b(n)\ ]\} \bmod 1 , \tag{15}$$

$$p_b(n+1) = \left\{ p_b(n) + \frac{k_b}{2\pi} \sin [\ 2\pi q_b(n)\ ] \right.$$
$$\left. - fb\ p_s(n+1)\ g'[q_s(n) + q_b(n)] \right\} \bmod 1 , \tag{16}$$

$$q_b(n+1) = \{q_b(n) + p_b(n+1)\} \bmod 1 \ , \tag{17}$$

with g given above.

We have varied the parameters b, f, $k_b$, and $k_s$ independently for a variety of initial conditions. First we will describe the behavior as the map parameters are varied, then we will report on the results of holding the parameters fixed and varying the initial conditions. We can summarize the behavior more or less as follows: for b = 0, the value of f is obviously unimportant, and the equations reduce to two, uncoupled, area preserving standard maps. The results presented here are mostly for b < 0 and |b| ≪ 1. The case b > 0 is even more complicated and the structure and position of the attractors varies more than the b < 0 case. The qualitative picture of what we find is shown in Figures 4 through 7. Details are given in the Figure captions. We exhibit only the projection of the attractor onto the $q_s$, $p_s$ plane. It is important to note the scale on the figures and recall that the full phase space lies in 0 to 1 for each degree of freedom. The size of the attractor is proportional to the magnitude of the coupling b, at least for |b| ≪ 1 and b < 0. For b > 0 the attractors are no longer confined to this small region of phase space, but typically exist around the period 2 fixed point of the standard map as shown in Figure 7. As we increase $k_b$, thus making the "bath" more and more chaotic, while holding $k_s$, b, and f fixed, we see the attractors change their nature from quasiperiodic to chaotic. The power spectrum for the last 65536 = $2^{16}$ points on the orbit is shown for two such examples. The dimension of the quasiperiodic attractors is presumably integer while that of the chaotic attractor is probably fractional. We have not computed

these dimensions, but infer these qualitative statements from the nature of

the power spectra. The broadband power spectrum of Figure 11 implies the

attractor is chaotic. Changing f alters the shape of the attractor but

little else. Varying $k_s$ has much the same effect as varying $k_b$.

Next we report on the features observed when the map parameters: $k_s$,

$k_b$, b, and f are held fixed and the initial conditions are varied. We find

that each initial condition we chose gave rise to a different attractor.

The map parameters were held fixed at f = 1.0, b = -0.01, $k_b$ = 1.74, and $k_s$

= 0.82 with $q_b(0)$ and $p_b(0)$ held at .3875 and 0.7654 respectively. Some of

the results are shown in Figures 13-19. These initial values of $p_b$ and $q_b$

define a plane in $p_s$-$q_s$ space in which we chose a sequence of order 150

different initial conditions separated by about 0.005 along the lines

indicated in Figure 12. We iterated the map 30,000 times for each initial

condition and threw away the first 10,000 points to let the transients die

away before plotting. Of the 150 initial conditions we tried, no two of

the attractors were exactly alike. This test indicates that the number of

attractors is very large, and we conjecture the number is in fact infinite.

We have performed several tests to check the results we have seen.

The first test we performed was to iterate the map a long number of times.

We iterated the map 1,000,000 times for a few different initial conditions

and found the attractor at this time was the same as at 100,000 iterations.

Each time we only kept the last 20,000 iterations to plot. We also

performed a series of experiments iterating the map for ~ 140,000 steps and

plotting the attractor every 40,000 time steps. We found the power

spectrum to be the same in every case where the transients died away before the start of the spectrum calculation. We iterated 10,000 times before we started calculating the power spectrum. If transients were still present, then the power spectrum would be different for the time series with the transients and the series without. The spectra were never identical, but all of the major peaks were reproduced to a high precision. The reason for the discrepency is presumably due to very long time scales required to fill out the attractor, and the Fourier transform algorithm we used. These tests convince us that the attractors we see are stationary. We have also experimented with changes in initial conditions as small as 1 part in $10^{-8}$ and found that these points also lead to different attractors. This is a further indication that we are actually seeing a continuum of attractors.

There may be some disagreement over calling the limit sets we observe attractors. We would like to emphasize the fact that globally, phase space volume is contracting. This is easily seen by starting with a hyper-cube of initial conditions which fills the entire phase space and see that this hyper-cube always gets attracted down onto a very small region of phase space near the limit sets we observe. It is for this reason we call the sets attractors. In another sense, the sets do not fit the conventional notion of an attractor since if we make a small perturbation of the orbit off of the limit set, the orbit will in general go to a new limit set. We choose to call these sets attractors with very complicated basin boundaries. We chose to use this terminology loosely instead of developing a whole new language for describing these sets.

We should make it clear that not every initial condition leads to a separate attractor. Consider the orbit generated by the initial condition $z(0)$ where $z(n) = (q_s(n), p_s(n), q_b(n), b_b(n))$. The orbit will be $z(0)$,

$z(1)$, $z(2)$, $z(3)$, ... $z(n)$, ..., where after some N the transients have died away and we say $z(N)$, $z(N+1)$, ... defines the attractor. Obviously we could also have chosen any one of these points as our initial condition, so each of these initial conditions lead to the same attractor. Hence not every initial condition can lead to a different attractor. This does not contradict the possibility of a continuum of attractors. It is possible that almost every initial condition in a plane or a cube yields a separate attractor. Consider a plane of initial conditions in the $p_s$ - $q_s$ plane defined by the point $p_b(0) = c1$, $q_b(0) = c2$. The set of points in the $p_s$ - $q_s$ plane that return to this plane under the map is the same set of points that will return to the values $p_b(0) = c1$, $q_b(0) = c2$. This is probably a set of measure zero since we know that the set of periodic orbits in the standard map is a set of measure zero for $k > 0$. Hence, almost every orbit in the plane is probably a separate orbit from any other in the plane and can thus yield separate attractors. For a cube, the cube would be defined by a line in $p_b$ - $q_b$ space and a plane in the $p_s$ - $q_s$ space. Again almost every point in the cube could yield a separate attractor.

Perhaps what we are seeing is a case where the basins of attraction are of smaller dimension than the original phase space. If that is indeed the case, then it would take an infinite number of basins of attraction to fill the entire phase space.

All of the attractors we have shown for this collection of initial conditions are quasiperiodic as can be seen from their power spectra. At higher values of either $k_s$ or $k_b$ the attractors become chaotic as mentioned earlier. We show the quasiperiodic attractors only because it is easier to pick out the differences from one attractor to another with one's own eye.

Another way to emphasize these differences is to look at the projections of the asymptotic orbit on other planes in the four dimensional phase space. Some of these projections are shown in Figures 20 to 23. Note that the projection on the $p_b$ - $q_b$ plane is very regular and close to the orbits seen in the uncoupled maps.

The attractors can be loosely classified into a few different types. One type is similar to that of Figure 4 which looks like the projection of the surface of a torus and forms a ring in the $p_s$ - $q_s$ plane. Still others form 3 or 5 or even higher numbers of rings in this plane. Other attractors such as that shown in Figure 1 exist in which the projection is even more complicated and intricate. One might suspect that all attractors of a given type are actually the same attractor. As far as we can tell, this is not the case. Even if one imagines filling out the attractors with more points, the attractors do not lie exactly on top of one-another. Thus each attractor is distinct.

This multitude of attractors has not been previously discussed in the literature to our knowledge. That may well be because, while it was known to be a possibility, no examples were known. From our preliminary investigations reported here, it would appear that we are observing an infinite number of attractors -- quasiperiodic and strange. Clearly this is a phenomenon quite dissimilar from the textbook picture of damped motion of a particle in a potential well with many local, isolated minima to represent the idea of several basins of attraction. If, indeed, we have so many attractors, the notion of basins of attraction capable of partitioning the phase space has disappeared altogether for dynamical systems such as we are dealing with here.

IV.  A Few Observations

Though a complete understanding of the phenomena revealed by our calculations has eluded us, we do have some observations about our map, Equations (14), (15), (16), (17), which give a qualitative hint as to their behavior.  First we note that there is a fixed point of the map in the neighborhood of the attractor.  A fixed point requires $p_b(n)$ to vanish, and it results in the values

$$k_b \sin [ 2\pi q_b ] = k_s \sin [ 2\pi q_s ] ,$$

(18)

$$p_s = - bg(q_s + q_b) ,$$

(19)

and

$$p_s = \frac{k_b \sin [ 2\pi q_b ]}{2\pi f b g'(q_s + q_b)}$$

(20)

for the other phase space coordinates.  For small b, we find a fixed point at approximately $q_s = 1/2$, $q_b = 1/2$ which implies $p_s = - b/2$.  For negative b, the fixed point resides near the center of the attractors and scales proportional to b just as the attractors in Figures 4-7 do.  For positive b the fixed point still lies near -b/2 but because we are mapping the variables back onto a torus, this point ends up at 1 - b/2.  Hence there is no fixed point in this region for positive b.  We have also observed no attractors in this region for b positive.  An obvious guess is that the attractors are associated with the connection between these facts.

We can easily evaluate the eigenvalues of the linearized mapping in the neighborhood of the indicated fixed point. We have done this for small b and find the values

$$\lambda_1 = \frac{\alpha + \sqrt{\alpha^2 - 4}}{2} \quad \text{and} \quad \lambda_2 = \frac{\beta + \sqrt{\beta^2 - 4}}{2} , \tag{21}$$

where

$$\alpha = 1 + b + \frac{1 - k_s}{1 + b} \quad \text{and} \quad \beta = 2 - k_b , \tag{22}$$

to leading order in b. Since the Jacobian of the map is unity in any local neighborhood of phase space and the mapping came from a canonical generating function, we find with no surprise the other eigenvalues to be

$$\lambda_3 = \frac{1}{\lambda_1} , \qquad \lambda_4 = \frac{1}{\lambda_2} . \tag{23}$$

The fixed point is thus elliptic for $k_b < 4$ and $k_s < 4(1+b)$. So we see that for that range of $k_s$ and $k_b$, into which most of the examples we have shown here fall, the attractor is associated with an elliptic fixed point. The usual standard map also has an elliptic fixed point at $q = 1/2$ and $p = 0$ for $k < 4$, however, the global area preservation property of that map results in far less structure than we have observed here. Perhaps the best remark we can make then is that while hints are given by the presence of the elliptic fixed point of the map, the global structure plays an essential role which we are presently unable to make clear.

This is perhaps the place to end our discussion. We have reported the features of an iterated map in four dimensions which locally has Jacobian unity, but, because we have chosen the phase space to lie on the torus $T^4$, it does not globally preserve phase space volume. This has a profound effect on the properties of the asymptotic behavior of the map. The final map has no particular connection to a dynamical system arising from a physical problem. In that sense it is an exploratory tool much like the Lorenz equations[5] or the Henon map.[4] The results of studying these tools is to identify generalizable properties of dynamical systems and to use the observations of those results as metaphors to describe qualitative behavior in realistic systems.

We have more recently looked at other maps which display similar behavior. We have looked at a few two dimensional maps which also seem to display an infinity of attractors. These studies add much support for the phenomena we are seeing in the 4 dimensional system. It seems this phenomena might exist in any system that is globally not invertible, but locally has a Jacobian of 1.

Our experiments on this map indicate that one can achieve an infinite number of attractors in a dynamical system. This is an important result even though established at this stage in a few systems and may be especially important in the interpretation of observed experimental phenomena in the behavior of nonlinear systems. Indeed, if a real physical system is found which has this property, then one would expect irreproducible asymptotic results for any experiment which is unable to precisely reproduce the initial conditions perfectly. We are familiar with the feature of chaotic systems that the detailed <u>orbits</u> are sensitive to initial conditions and are, in general, irreproducible experimentally. That the <u>asymptotic attractors</u> should also be uncertain is a new item to contemplate.

Finally, a conjectured interpretation of our results is that we are observing a system with basins of attraction whose dimension is less than that of the full phase space. Such basins of attraction would not partition the phase space.

**Figure 1:** The projection of the orbit onto the $p_s$-$q_s$ plane for the iterated map equations (14-17) of the text for parameter values b = -.1, f = 1.0, $k_s$ = 1.74, $k_b$ = 0.82. 13,000 iterations of the map were taken starting at the initial conditions $p_s(0)$ = 0.2, $q_s(0)$ = 0.8, $p_b(0)$ = 0.5234, $q_b$ = 0.05. In the figure iterations 5,000 to 13,000 are shown. The transients died away in the first 5,000 steps. Note the difference between this and figures 2 and 3.

**Figure 2:** Same as figure 1, but with b = -.01, $k_s$ = 2.01, $q_s(0)$ = 0.2. All other parameters and time steps are the same. Note the totally different structure of this attractor compared with figures 1 and 3.

**Figure 3:** Same as figure 1, except $k_s$ = 0.323, $k_b$ = 0.3131, and b = -0.01. All other parameters and iterations are the same.

**Figures 4-7:** In this set of pictures, b is varied as all the other parameters and the initial conditions are held fixed at $k_s$ = 0.82, $k_b$ = 0.74, f = 1.0, $p_s(0)$ = 0.26, $q_s(0)$ = 0.9, $p_b(0)$ = 0.3875, $q_b$ = 0.7654, b = -0.1 in figure 4; b = -0.01 in figure 5; b = -0.005 in figure 6, and b = +0.1 in figure 7. Note the scales on the plots and how the attractor scales with b for b negative. Also note that for b positive, the attractor is in a different region of phase space. For b > 0 the attractor is typically near the period 2 fixed point of the standard map, whereas for b < 0 the attractor is near the period 1 fixed point.

**Figure 8:** The projection of the orbit onto the $p_s$-$q_s$ plane with parameter values $k_s$ = 0.74, $k_b$ = 2.74, b = -.02, f = 5.13. 141,536 iterations of the map were taken starting at the initial conditions $p_s(0)$ = 0.26, $q_s(0)$ = 0.65, $p_b(0)$ = 0.3875, $q_b(0)$ = 0.7654. In the figure iterations 85,000 to 95,000 are shown. This is one of a series of tests done to make sure the attractor was invariant. Every 20,000 iterations were checked to see if the attractor changed its shape. The appearance of the attractor remained the same in each block of the time series. Usually all transient behavior died away within the first 5,000 iterations.

**Figure 9:** The power spectrum of the time series of $p_s$ of the attractor shown in figure 8. The power spectrum was calculated on the last 65,536 ($2^{16}$) iterations. Notice there are only a small number of frequencies present indicating the quasiperiodic nature of the attractor.

**Figure 10:** The same as figure 8, but we have increased $k_b$ to 9.485 which is well into the chaotic regieme. The attractor no longer has the quasiperiodic structure shown in figure 8. Note the scale on the picture. So that the final set of points fits nicely on one small region, values of $p_s$ near 1 were plotted as $p_s$-1, otherwise part of the attractor would not be resolved in the picture.

**Figure 11:** The power spectrum of the time series of $p_s$ in figure 10. The same thing was done in this plot as in figure 9. The broadband power spectrum indicates the attractor is chaotic.

**Figure 12:** The initial conditions for checking the number of independent attractors. We choose approximately 150 initial conditions as shown in the figure, and held all the parameters, $p_b(0)$, and $q_b(0)$ fixed. In this test, we found no two attractors that were identical. The $q_s$ initial conditions were incremented by 0.005 on the lower parabola and by 0.01 on the two horizontal lines. Tests were also done with points as close as one part in 100,000,000 and it was found that these initial conditions also lead to different attractors. This experiemnt shows that there are at least 150 separate attractors in this system. This fact does not prove the existence of an infinite number of attractors, but it is a good indication.

**Figures 13-19:** Some of the attractors generated from the initial conditions of figure 12. In each of these pictures the parameters were held fixed at $k = 1.74$, $k = .82$, $b = -.01$, $f = 1.0$, $p_b(0) = .7654$, $q_b(0) = 0.3875$. $p_s(0)$ and $q_s(0)$ vary from picture to picture as indicated in each.

**Figures 20-23:** An attractor is projected onto 4 different planes. The parameter values for the map are $k_s = .82$, $k_b = 1.75$, $b = -.01$, $f = 5.0$. Initial conditions are $q_s(0) = .2$, $p_s(0) = .6$, $q_b(0) = 0.05$, and $p_b(0) = 0.5234$. Notice the orbit projected onto the $p_b$-$q_b$ plane infigure 21 is very close to an orbit of the uncoupled standard map. Also notice how the quasiperiodic structure of the attractor is shown clearly in figures 22 and 23.

188

References

[1] A. J. Lichtenberg and M. A. Lieberman, Regular and Stochastic Motion, Springer, New York (1983).

[2] J. Guckenheimer and P. Holmes, Nonlinear Oscillations, Dynamical Systems, and Bifurcations of Vector Fields, Springer, New York (1983).

[3] Dynamical Systems and Chaos, Proceedings of the Sitges Conference, Springer Lecture Notes in Physics (1982).

[4] M. Hénon, Commun Math. Phys. 50, 69 (1976).

[5] E. N. Lorenz, J. Atmos. Sci. 20, 130 (1963).

[6] J. D. Farmer, E. Ott and J. A. Yorke, Physica 7D, 153 (1983).

[7] P. Grassberger and I. Procaccia, Phys. Rev. Lett. 50, 346 (1983).

[8] A. Brandstater, et al., Phys. Rev. Lett. 51, 1442 (1983).

[9] A. Ben-Mizrachi, I. Procaccia, and P. Grassberger, Phys. Rev. A 29, 975 (1984).

[10] R. Shaw, Z. Naturforsch, 36 A, 80 (1981); H. D. I. Abarbanel and P. E. Latham, Phys. Lett. 89 A, 55 (1982).

[11] C. Grebogi, S. W. McDonald, E. Ott, and J. A. Yorke, Phys. Lett. 99 A, 415 (1983).

[12] S. W. McDonald, C. Grebogi, E. Ott, and J. A. Yorke, Fractal Basin Boundaries, Plasma Preprint UMLPF #84-017 (1984).

[13] R. H. G. Helleman, in Statistical Mechanics, Vol. 5, edited by E. G. D. Cohen (No. Holland. Publ., Amsterdam, 165, 1980).

[14] N. R. Wagner, Proc. of Symp. on Security & Privacy, Oakland, CA. Apr. 29 - May 2 1984 (IEEE Comp. Soc. Press) 91.

[15] R. S. Winternitz, Proc. of Symp. on Security & Privacy, Oakland, CA. Apr. 29 - May 2 1984 (IEEE Comp. Soc. Press) 88-90.

[16] B. A. Huberman, Computing with Attractors -- to appear in The Emergence of New Synthesis in Science.

[17] T. Hogg and B. A. Huberman, Attractors on Finite Sets: The Dissipative Dynamics of Computing Structures (Preprint).

[18] J. J. Hopfield, Proc. Natn'l Acad. Sci. 79, 1554 (1982).

[19] B. A. Huberman and T. Hogg, Phys. Rev. Lett. 52, 1048 (1984).

[20] M. R. Guevara, L. Glass, M. C. Mackey, and A. Shrier, Chaos in Neurobiology, IEEE Trans. SMC 13, 790 (1983).

[21] N. B. Abraham, J. P. Gollub, and H. L. Swinney, Testing Nonlinear Dynamics, Physica 11 D, 253 (1984).

[22] B. V. Chirikov, Phys. Rep. 52, 263 (1979).

[23] B. V. Chirikov, Sov. J. Plasma Phys. 5, 263 (1979a).

[24] J. M. Greene, J. Math. Phys. 20, 1183 (1979).

[25] R. S. Mackay, Princeton University Thesis, University Microfilm International, Ann Arbor, MI (order #830 1141, 1982).

[26] H. Goldstein, Classical Mechanics, 2nd ed., p. 378 (Addison Wesley, London, 1980).

[27] A. J. Lichtenberg and M. A. Lieberman, Regular and Stochastic Motion, Springer, New York, p. 224 (1983).

[28] F. Vivaldi,,Rev. Mod. Phys. 56, 737 (1984).

[29] C. Froeschlę, Astrophys. Space Sci. 14, 110 (1971).

[30] C. Froeschle, Astron. Astrophys. 16, 172 (1971).

[31] R. Bagley and K. Kaneko, Phys. Lett. NN. PP. (1985).

FIGURE 1

# P s   v s .   Q s

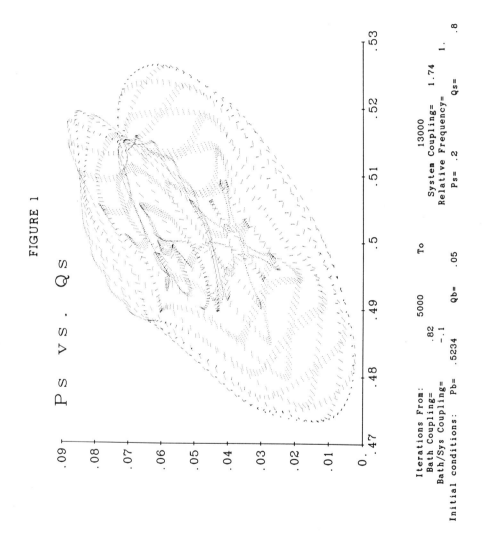

Iterations From:        5000        To        13000
        Bath Coupling=        .82                System Coupling=        1.74
        Bath/Sys Coupling=        -.1                Relative Frequency=        1.
Initial conditions:        Pb=        .5234        Qb=        .05        Ps=        .2        Qs=        .8

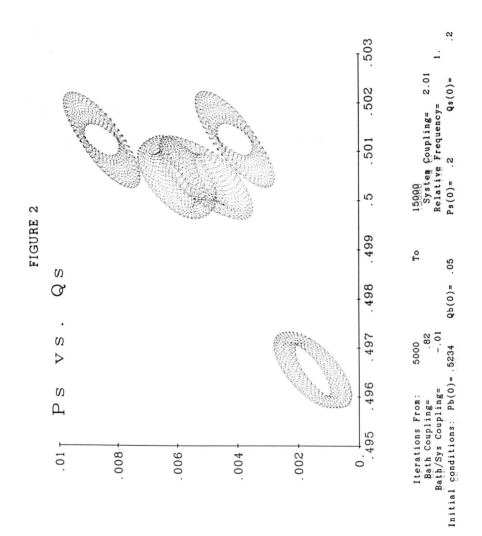

FIGURE 2

P s   v s .   Q s

Iterations From:    5000           To          15000
       Bath Coupling=       .82              System Coupling=    2.01
  Bath/Sys Coupling=      -.01              Relative Frequency=     1.
Initial conditions:  Pb(0)= .5234      Qb(0)= .05      Ps(0)=  .2      Qs(0)=  .2

FIGURE 3

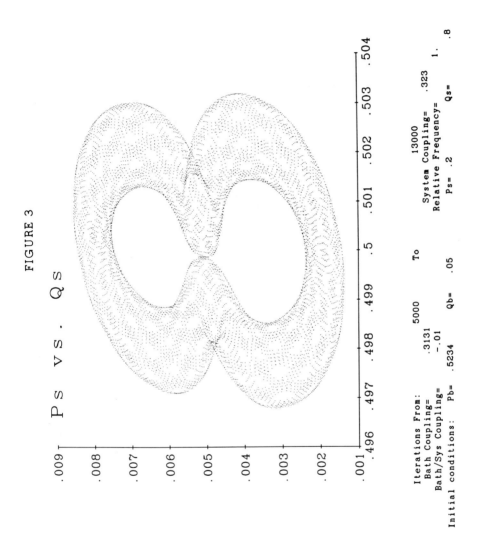

P s   v s .   Q s

Iterations From:                    5000                To              13000
        Bath Coupling=              .3131                    System Coupling=    .323
        Bath/Sys Coupling=         -.01                     Relative Frequency=    1.
Initial conditions:   Pb=  .5234         Qb=  .05             Ps=  .2         Qs=  .8

FIGURE 4

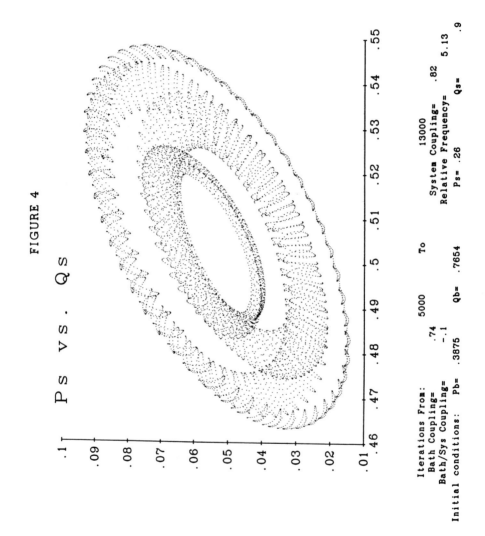

P s  v s .  Q s

Iterations From:           5000        To        13000
    Bath Coupling=      .74                  System Coupling=      .82
    Bath/Sys Coupling=    -.1              Relative Frequency=    5.13
Initial conditions:   Pb=  .3875      Qb=  .7654        Ps=  .26        Qs=    .9

193

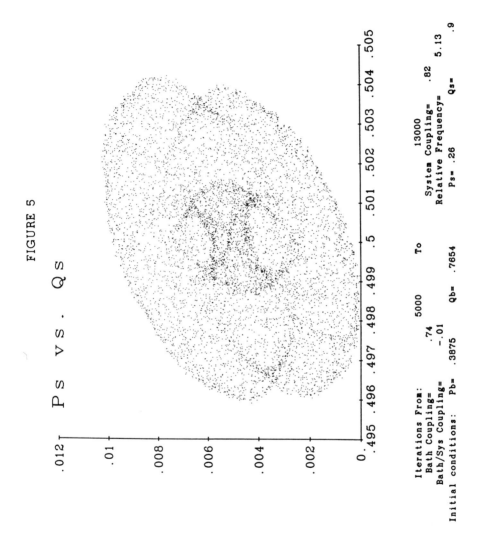

FIGURE 5

Ps vs. Qs

194

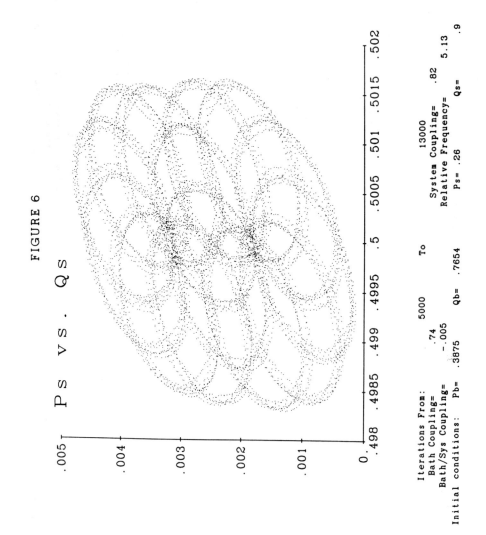

FIGURE 6

P s  v s .  Q s

Iterations From:        5000        To        13000
    Bath Coupling=        .74              System Coupling=        .82
    Bath/Sys Coupling=   -.005            Relative Frequency=    5.13
Initial conditions:   Pb=  .3875   Qb=  .7654        Ps=  .26        Qs=  .9

FIGURE 7

Pb vs. Qb

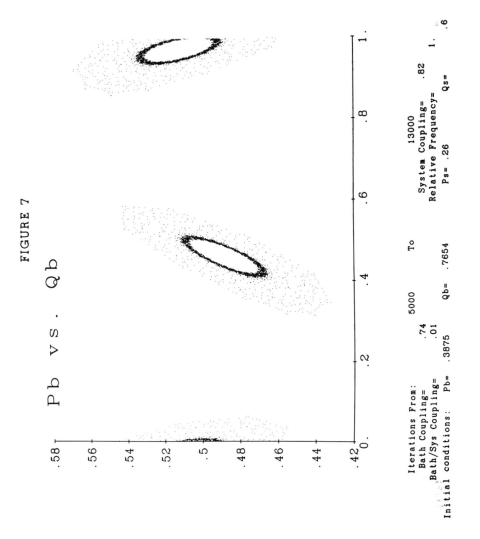

Iterations From:                                        To
        Bath Coupling=            .74                        System Coupling=      13000
    Bath/Sys Coupling=     .01            5000        Relative Frequency=          .82          1.
Initial conditions:    Pb=    .3875            Qb=    .7654          Ps=    .26        Qs=          .6

FIGURE 8

Ps vs. Qs

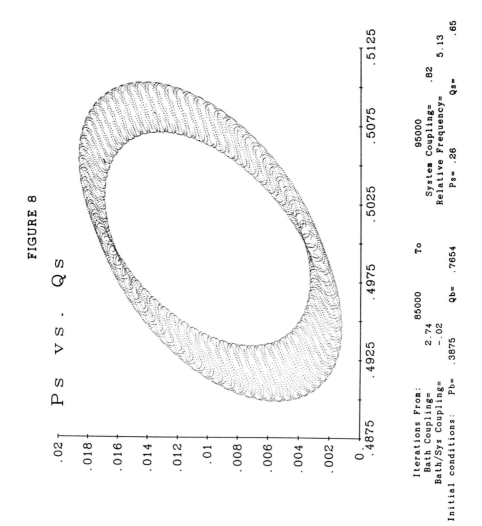

Iterations From:       85000        To           95000
Bath Coupling=         2.74              System Coupling=    .82
Bath/Sys Coupling=     -.02              Relative Frequency=  5.13
Initial conditions:    Pb=  .3875    Qb=  .7654    Ps=  .26    Qs=  .65

.02
.018
.016
.014
.012
.01
.008
.006
.004
.002
0

.4875   .4925   .4975   .5025   .5075   .5125

FIGURE 9

Ps  Power  Spectrum

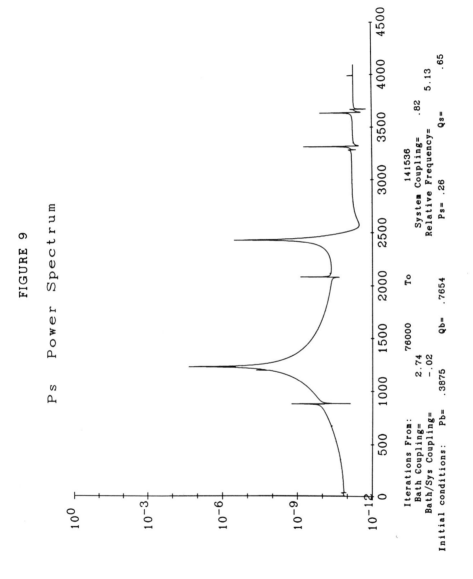

Iterations From:              76000          To            141536
    Bath Coupling=        2.74                      System Coupling=        .82
    Bath/Sys Coupling=   -.02                      Relative Frequency=       5.13
Initial conditions:    Pb= .3875      Qb= .7654           Ps= .26        Qs=      .65

FIGURE 10

P s   v s .   Q s

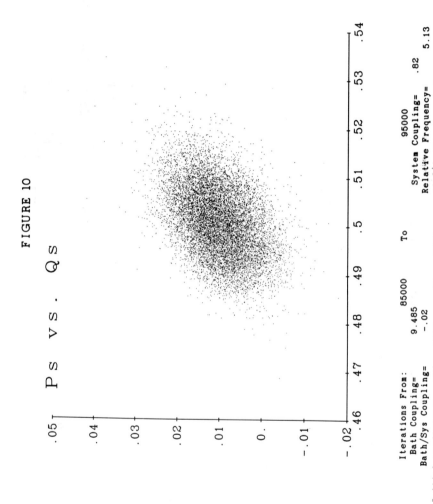

Iterations From:            85000            To            95000
         Bath Coupling=      9.485                 System Coupling=      .82
    Bath/Sys Coupling=       -.02            Relative Frequency=     5.13
Initial conditions:    Pb=  .3875      Qb=  .7654        Ps=  .26        Qs=      .65

FIGURE 11

Ps Power Spectrum

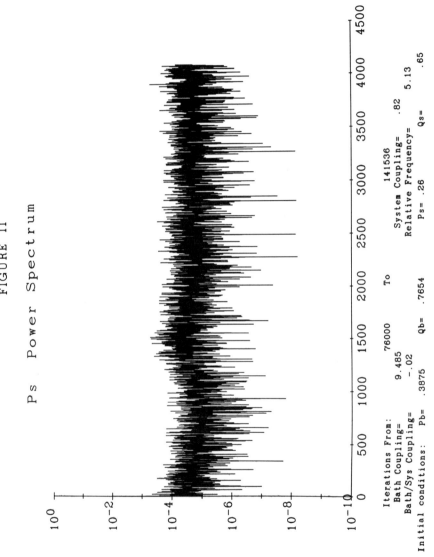

Iterations From:        76000          To        141536
     Bath Coupling=     9.485                System Coupling=    .82
Bath/Sys Coupling=      -.02                Relative Frequency=    5.13
Initial conditions:  Pb=  .3875    Qb=  .7654      Ps=  .26    Qs=   .65

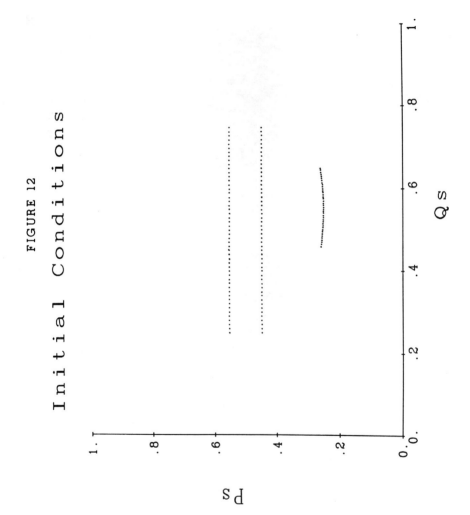

FIGURE 12

Initial Conditions

FIGURE 13

P s   v s .   Q s

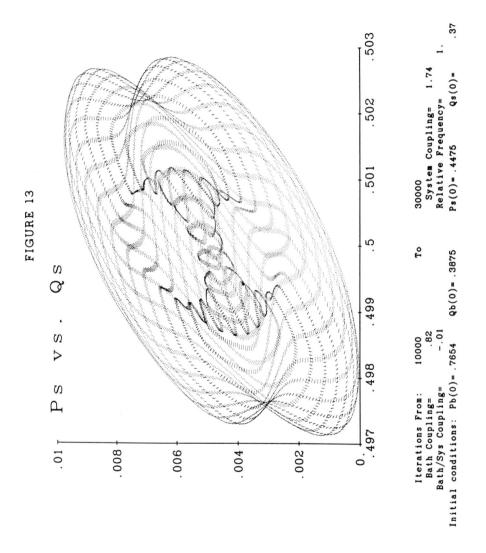

| | | | |
|---|---|---|---|
| Iterations From: | 10000 | To | 30000 |
| Bath Coupling= | .82 | | System Coupling= 1.74 |
| Bath/Sys Coupling= | -.01 | | Relative Frequency= 1. |
| Initial conditions: Pb(0)= .7654 | Qb(0)= .3875 | Ps(0)= .4475 | Qs(0)= .37 |

202

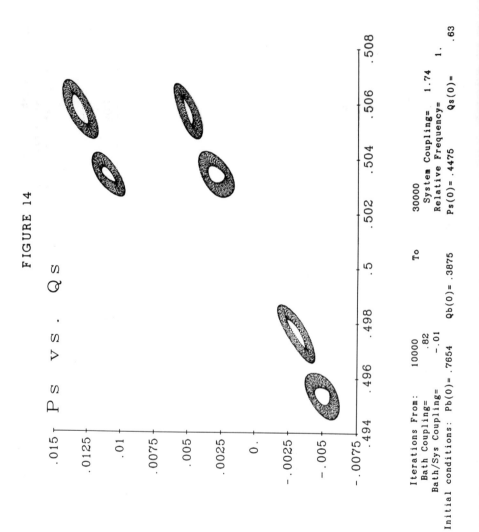

FIGURE 14

FIGURE 15

## Ps vs . Qs

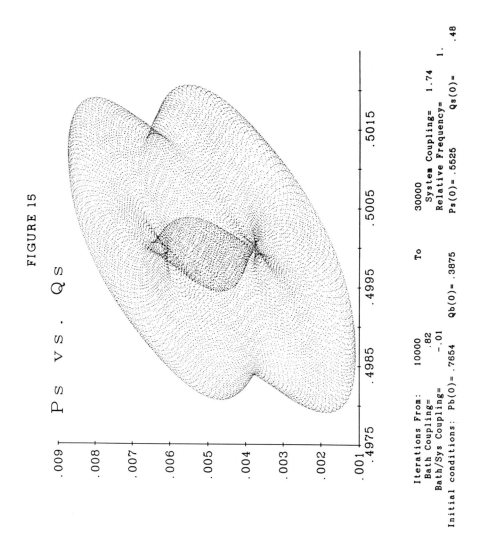

| Iterations From: | 10000 | To | 30000 | |
|---|---|---|---|---|
| Bath Coupling= | .82 | | System Coupling= | 1.74 |
| Bath/Sys Coupling= | -.01 | | Relative Frequency= | 1. |
| Initial conditions: | Pb(0)= .7654 | Qb(0)= .3875 | Ps(0)= .5525 | Qs(0)= .48 |

204

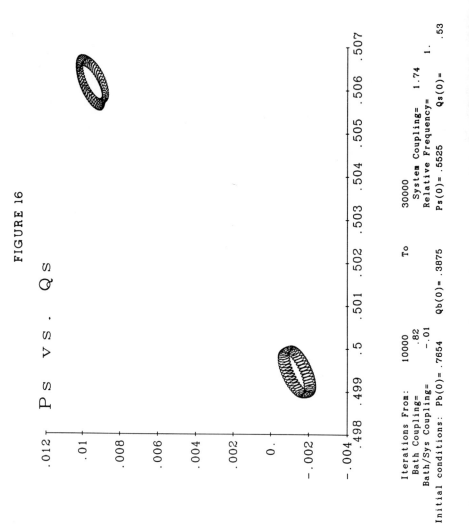

FIGURE 16

Ps vs. Qs

Iterations From:    10000           To        30000
        Bath Coupling=     .82           System Coupling=    1.74
    Bath/Sys Coupling=    -.01          Relative Frequency=   1.
Initial conditions: Pb(0)= .7654    Qb(0)= .3875    Ps(0)= .5525    Qs(0)=    .53

205

FIGURE 17

P s   v s .   Q s

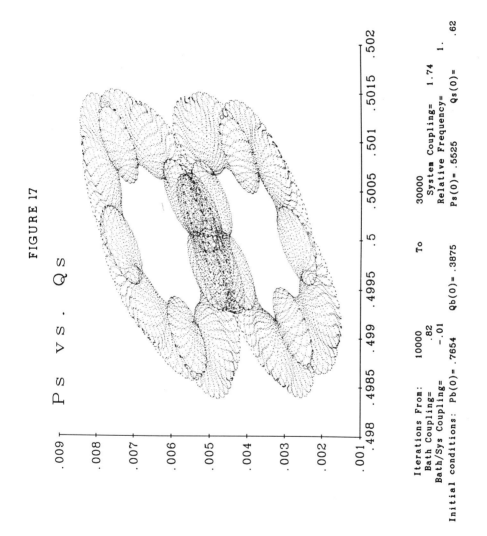

| Iterations From: | 10000 | To | 30000 | | |
| Bath Coupling= | .82 | | System Coupling= | 1.74 | |
| Bath/Sys Coupling= | -.01 | | Relative Frequency= | | 1. |
| Initial conditions: | Pb(0)= .7654 | Qb(0)= .3875 | Ps(0)= .5525 | Qs(0)= | .62 |

FIGURE 18

# P s  v s .  Q s

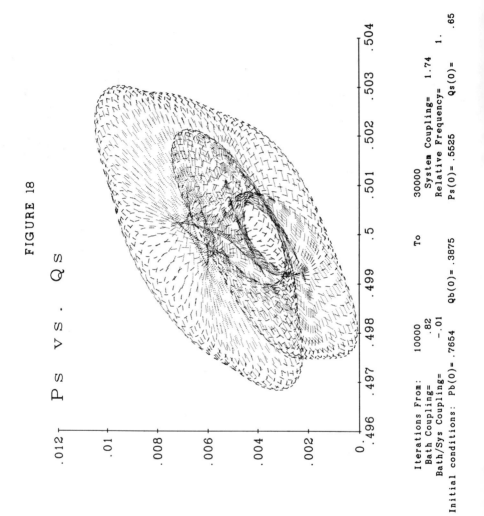

Iterations From:     10000          To          30000
       Bath Coupling=        .82              System Coupling=    1.74
   Bath/Sys Coupling=       -.01          Relative Frequency=    1.
Initial conditions:  Pb(0)= .7654     Qb(0)= .3875     Ps(0)= .5525     Qs(0)=    .65

FIGURE 19

## P s   v s .   Q s

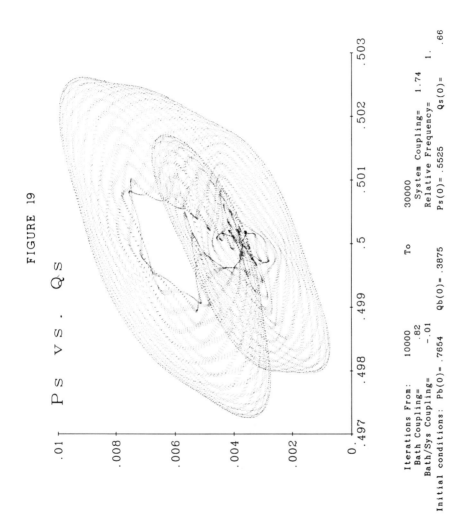

| Iterations From: | 10000 | | To | 30000 | |
|---|---|---|---|---|---|
| Bath Coupling= | .82 | | | System Coupling= | 1.74 |
| Bath/Sys Coupling= | -.01 | | | Relative Frequency= | 1. |
| Initial conditions: | Pb(0)= .7654 | Qb(0)= .3875 | | Ps(0)= .5525 | Qs(0)= .66 |

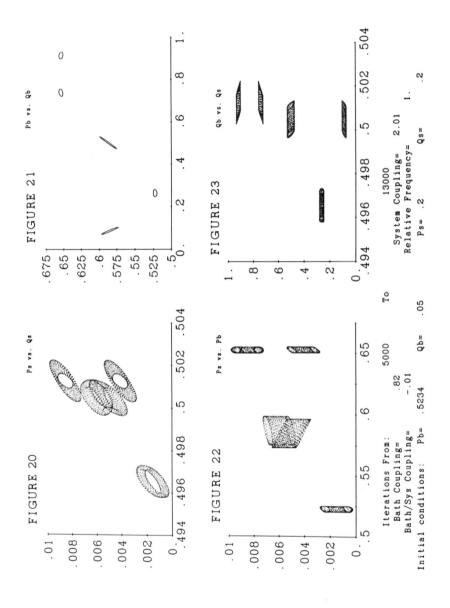

FIGURE 20

Ps vs. Qs

FIGURE 21

Pb vs. Qb

FIGURE 22

Ps vs. Pb

FIGURE 23

Qb vs. Qs

Iterations From:     5000     To     13000
Bath Coupling=        .82           System Coupling=    2.01
Bath/Sys Coupling=    -.01          Relative Frequency=  1.
Initial conditions:  Pb=  .5234     Qb=  .05     Ps=  .2     Qs=  .2

# NONLINEAR DYNAMICS IN OPTICAL SYSTEMS

N. B. Abraham

Department of Physics, Bryn Mawr College,

Bryn Mawr, PA 19010 USA

## ABSTRACT

Optical systems provide a wealth of nonlinear dynamical phenomena with a wide range of easily measurable temporal and spectral characteristics and with a variety of technological applications.

## 1. Introduction

The history of studies of nonlinear dynamics in lasers and other optical systems parallels that of the more classic studies of chaos and nonlinear dynamics, with the earliest puzzling about periodic and chaotic pulsations in lasers and masers coming in the late 1950's and early 1960's [1,2]. It was not until more than ten years later [3] that it was realized that the fundamental equations used to describe the laser dynamics were essentially equivalent to those used in the classic Lorenz model [4]. Since then the study of nonlinear dynamics in optical systems as a fundamental investigation of chaotic and irregular behavior has become as important as the ever growing interest in technological applications of optical components and devices. An excellent survey of the status of nonlinear dynamics in optical systems can be gained from several key compilations in special issues of journals [5,6], in conference proceedings [7-9], in soon to be published books [10,11] and in the multitude of references cited in these sources. In addition, there seems to be a steady continuation of conferences on this subject which may attract and lead to publication of the latest new results [12,13]. The generic subfields in optics are "laser instabilities" and "optical bistability" with such related topics as "instabilities", "spontaneous pulsations", and "optical chaos". It is worth noting that both intrinsic and hybrid nonlinear electro-optical systems which have multivalued steady state solutions (generically, though imprecisely, called "bistable systems") have shown a wealth of pulsations so that "optical bistability" is often colloquially used to cover dynamical pulsations as well as truly bistable systems.

## 2. The special advantages of optical systems

The details of various optical systems can be garnered from the appropriate specific articles. It is important to note here that the intrinsic pulsation frequencies in optical systems have ranged from kilohertz to gigahertz and that periodic, period-doubled, quasiperiodic, intermittent, and chaotic signals of various types and sequences have been observed. Unstable optical systems have included conventional laser devices, modulated systems, and those with external feedback loops. When the feedback has been slow with respect to the response time of the medium (as is possible in electro-optic hybrids), the systems have provided excellent testing grounds for the validity of one-dimensional maps as descriptors of physical behavior. It is also true that optical systems have provided the best (and perhaps only) physical realization of the Lorenz equations [14]. Recent reports show evidence for periodic and chaotic behavior as predicted in theoretical analyses using the appropriate parameters [15,16].

Optical systems have provided signals suitable for analysis of fractal dimensionality of the attractors, calculation of metric or information entropies, and determination of Lyapunov exponents. The frequency range of the pulsations is appropriate for data acquisition with fast electronics and is ideally suited for avoidance of environmental noise. There is some hope that the fastest pulsation rates may provide opportunities for real-time analog processing (perhaps by nonlinear optical means) that will give quick evidence of the underlying dynamical origin of pulsations. At the same time, optical systems often come with the intrinsic spontaneous emission noise, so we are able to test the stability and robustness of the semiclassical models while looking for the effects of noise perturbations and quantum noise disturbances in particular.

In addition, it is quite a common matter to examine optical signals for temporal and/or spatial coherence. It is not presently clear how traditional measures of coherence or some of the rarely used correlation measures might be helpful in characterizing signals and in determining the underlying physical processes. An area of quite active concern, and certainly an area of growing research interest, is that of the coherence of chaotic signals, and one may expect some considerable progress in the near future from optical physicists who are trained in classical or quantum coherence theory and measurements. We may also soon find that spatial pattern formation and variation within optical systems will modulate the spatial and temporal coherence of the optical output, giving us a direct measure of the underlying processes.

## 3. Technical Applications

The study of nonlinear dynamics is of special interest in a wide range of applications. Thus, phenomena such as mode hopping and mode locking have important device implications, and fundamental limits on the performance of a ring gyroscope or on a modulated laser diode used in communications may be given by these studies. High power and high resolution laser devices are susceptible to perturbations and may often press the dynamical limits of stable operation. The demand for faster optical switching devices and wider bandwidth optical communications hardware has accelerated the study of how to control or limit chaotic modes of operation. Interest in new forms of displays promotes consideration of spatial pattern formation and the regularities and irregularities in such systems. It is rather remarkable that almost every practical experiment reported in nonlinear dynamics has an analog in optical systems and is equally likely to find a technological application as well.

## 4. Some special examples

As an example of the many different forms of study, we include in the following article some specific recent results of our studies of single-mode ring lasers and our search for definite distinctions between chaotic and stochastic origins for observed laser output pulsations. Other examples can be found in References 5-11.

## 5. References

1. A. G. Gurtovnick, Izv. Vyss. Uchebn. Zaved. Radiofiz, **1**, 83 (1958); A. N. Oraevskii, Radio Eng. Electron. Phys. (USSR) **4**, 718 (1959).
2. A. V. Uspenskii, Rad. Eng. Electron. Phys. (USSR) **8**, 1145 (1963); **9**, 605 (1964); V. V. Korobkin and A. V. Uspenskii, Sov. Phys. JETP **18**, 693 (1964); A. Z. Grasyuk and A. N. Oraveskii in *Quantum Electronics and Coherent Light*, P. A. Miles, ed (Academic, New York, 1964), p.

192; Radio Eng. Electron. Phys. (USSR) **9**, 424 (1964); E. R. Buley and F. W. Cummings, Phys. Rev. **134**, A1454 (1964); H. Risken and K. Nummedal, J. Appl. Phys. **39**, 4662 (1968).

3. H. Haken, Phys. Lett. **53A**, 77 (1975).

4. E. N. Lorenz, J. Atmos. Sci **20**, 130 (1963).

5. J. Opt. Soc. Am. B, **2**, January 1985, Special Issue on Instabilities in Active Optical Systems, ed. N. B. Abraham, L. A. Lugiato, and L. M. Narducci.

6. IEEE J. Quantum Electron., **QE-21**, September 1985, Special Issue on Optical Bistability, ed. E. Gamire.

7. *Coherence and Quantum Optics V,* Proceedings of the Rochester Conference on Coherence and Quantum Optics, June 1983, ed. L. Mandel and E. Wolf (Plenum, New York, 1983).

8. *Optical Bistability 2,* Proceedings of the OSA Topical Meeting on Optical Bistability, June 1983, ed. C. R. Bowden, H. M. Gibbs, and S. M. McCall (Plenum, New York, 1983).

9. *Instabilities and Dynamics in Lasers and Other Nonlinear Optical Systems,* Proceedings of the International Meeting on Instabilities and Dynamics in Lasers and Other Nonlinear Optical Systems, June 1985, ed. R. W. Boyd, L. M. Narducci, and M. G. Raymer (Cambridge U. Press, Cambridge 1985).

10. *Instabilities and Chaos in Quantum Optics,* ed. by F. T. Arecchi and R. G. Harrison (Springer-Verlag, Heidelberg 1985/6).

11. *Instabilities and Chaos in Lasers, Lasers with Saturable Absorbers and Lasers with Injected Signals,* N. B. Abraham, P. Mandel, and L. M. Narducci, *Progress in Optics,* ed. E. Wolf, (North Holland, Amsterdam, to be published).

12. OSA Topical Meeting, **Optical Bistability 3**, Tucson, AZ, December 1985. (Chairmen: H. M. Gibbs, S. D. Smith, P. Mandel, and N. Peyghambarian)

13. SPIE meeting on **Optical Chaos** in the 1986 Quebec International Symposium on Topics in Optical and Optoelectronic Applied Sciences and Engineering, Quebec, Canada, June 1986. (Chairmen: J. Chrostowski and N. B. Abraham)

14. E. Hogenboom, W. Klische, C. O. Weiss, and A. Godone, Phys. Rev. Lett., (to be published); and in Reference 9.

15. L. W. Casperson, J. Opt. Soc. Am. B. **2**, 993 (1985).

16. L. M. Narducci, H. Sadiky, L. A. Lugiato, and N. B. Abraham, Opt. Commun. **55**, 370 (1985).

# Low-Dimensional Attractor for a Single-Mode Xe-He Ring Laser

N. B. Abraham, A. M. Albano, G. C. de Guzmán, M. F. H. Tarroja, and S. Yong
Department of Physics, Bryn Mawr College, Bryn Mawr, PA 19010

S. P. Adams and R. S. Gioggia
Department of Physics, Widener University, Chester, PA 19013

## Abstract

Calculations of the order-2 information dimension and the order-2 Kolmogoroff entropy of output intensity time sequences from an unstable, unidirectional, single-mode, inhomogeneously-broadened Xe-He ring laser indicate that the laser's dynamics is characterized by an attractor of relatively low dimensionality. This behavior, which is similar to that found in an earlier analysis of a Fabry-Perot laser using the same lasing medium, is deemed rather surprising in view of the fact that inhomogeneously-broadened lasers are usually modelled in phase spaces of very high dimensionality. The calculations were performed on relatively small, "noisy" data sets. Some evidence is presented which indicates that nontrivial dynamical information is attainable in spite of these limitations.

# Introduction

In recent years, it has become possible to describe the asymptotic time evolution of many experimental systems in terms of motions on strange attractors.[1-2] As techniques for calculating such parameters as dimensions, entropies and Lyapunov exponents from time sequence measurements have become better defined , the description of the behavior of experimental systems has evolved from qualitative classifications based on snapshots of time sequences or on power spectra to more quantitative descriptions using these calculated characteristics. While spectral studies have made possible the identification of such features of the dynamics as subharmonic bifurcations, mode-locking, or transitions to apparently chaotic behavior, they have not led to an unambiguous discrimination between signals resulting from "deterministic chaos" and those that are merely random. Dimensions, entropies and Lyapunov exponents make such a discrimination possible. They also provide a quantitative means of following the system as it evolves from one kind of behavior to another. In addition, the dimensions defined for an attractor provide estimates of the number of independent variables that may ultimately be needed to model the system when it is moving on that attractor.

In a recent paper [2], we reported on dimension and entropy calculations performed on digitized, output intensity time sequences from an unstable, inhomogenously-broadened, Fabry-Perot Xe-He laser. The results of the calculations indicated that for certain operating conditions, the temporal behavior of the output intensity could be characterized by a chaotic attractor of relatively low dimensionality.

The calculation of dimensions, entropies or Lyapunov exponents from experimental data is handicapped, however, by the presence of noise,whether inherent in the system or contributed by measuring instruments, and by technical limitations on the number of observations that can be made in single, uninterrupted sequences of measurements. For a Fabry-Perot laser, these limitations are compounded by the presence of counter-propagating fields in the cavity which lead to more complicated dynamical evolution.

The complications introduced by counter-propagating fields are eliminated by a unidirectional ring laser configuration. Previous work [4-5] on the ring laser system used in this study revealed the existence of a second threshold above the threshold for cw lasing at which spontaneous self-pulsing was observed. Depending on operating conditions, the spectra of the laser output intensity beyond this second threshold showed qualitative features characteristic of pulsations at a single frequency (period 1), or at two frequencies in the ratio 2:1 (period 2) or of chaotic or random behavior, among others. The preliminary results reported in this paper confirm that the broadband spectra observed at some operating conditions were indeed due to deterministic chaos rather than random noise. Taken together with the results obtained from the Fabry-Perot laser, these new results confirm that the dynamical evolution of this type of laser system is characterized by an attractor of low dimensionality in spite of the fact that the operation of these lasers have so far been modelled accurately only by a very large number of coupled nonlinear differential equations. [6]

## Experimental System

The laser system studied is described elsewhere[4] and shown schematically in Figure 1. It uses the high-gain 3.51 μm laser transition in xenon. The population inversion is created by dc discharges in the laser tubes, excited by well-regulated power supplies with series ballast resistors to control plasma instabilities. The cavity length can be controlled by moving one mirror with a piezo-electric crystal ( PZT) at one corner of the rectangle, and the system is made unidirectional by Faraday rotators (FR) and linear polarizers (P).

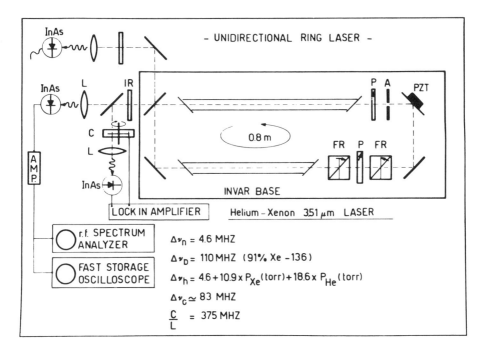

Figure 1. The ring laser configuration. The mirror in the upper right-hand corner is mounted on a piezoelectric transducer (PZT). Faraday rotators are labelled FR, and linear polarizers are labelled P.

For fixed pressure and discharge current, data sets were taken for different cavity detunings (or cavity lengths), which gave qualitatively different power spectra.

The intensity output was digitized by a Tektronix fast transient digitizer, which provided 10-bit resolution and sequences of up to 512 data points taken in intervals of $\tau = 1$ ns. The data sets were stored by an Apple microcomputer and were transferred to a Burroughs 6900 computer for analysis.

## Analysis

The order-2 information dimension, $D_2$, and order-2 Kolmogoroff entropy, $K_2$, were calculated for each data set using techniques developed by Grassberger and Procaccia:[3]

For each set of N numbers,

$$\{x_i \mid i = 1, 2, ..., N\},$$

where $x_j = x(j\tau)$, j an integer and $\tau$ the time interval between measurements, we formed n-dimensional time-delay vectors,

$$Y_k = (x_k, x_{k+1}, ..., x_{k+n-1}).$$

These vectors are used to evaluate the correlation sum,

$$C_n(\epsilon) = (1/N_n) \sum_{j \neq k} \Theta(\epsilon - |\mathbf{Y}_k - \mathbf{Y}_j|) \, ,$$

where $\Theta(...)$ is the Heaviside function, and $N_n$ is the number of pairs of n-dimensional vectors used in the sum. For sufficiently small $\epsilon$'s and large embedding dimensions, n, the correlation sum scales as,

$$C_n(\epsilon) \sim \epsilon^{D2} e^{-\tau n K2} \, .$$

Treating $C_n(\epsilon)$ as a function both of $\epsilon$ and the embedding dimension, n, $D_2$ and $K_2$ are obtained from log-log plots of $C_n(\epsilon)$ vs. $\epsilon$ - i.e.,

$$D_2 = \lim_{\epsilon \to 0} \lim_{n \to \infty} d[\log C_n(\epsilon)] / d[\log \epsilon] \, ,$$

and,

$$K_2 = \lim_{\epsilon \to 0} \lim_{n \to \infty} \log[C_n(\epsilon)/C_{n+1}(\epsilon)] \, .$$

In practice, $\epsilon$ is limited from below by noise and n is eventually limited from above by the size of the data set, although a more stringent limit on n is usually set by constraints on computing time.

For the purposes of discriminating between chaotic signals and random noise, the important properties of $D_2$ are that it should be 1 for a periodic signal, 2 for signals characterized by two incommensurate frequencies, finite and greater than 2 for chaotic signals and infinite for random signals. On the other hand, $K_2$ should be zero for periodic signals, finite but nonzero for chaos, and infinite for random noise.[3]

## Results

To estimate the extent of such limitations as are imposed on calculated dimensions by the smallness of experimental data sets , we performed the dimension calculations described above on sets of 500, 1200, 4000 and 10,000 points calculated from the Hénon map,

$$(x_{n+1}, y_{n+1}) = (y_n + 1 - ax_n^2, bx_n),$$

with a = 1.4 and b = 0.3. Figure 2 shows plots of $\log C_n(\epsilon)$ vs. $\log \epsilon$ for these data sets, all for embedding dimension = 4. The curves have been displaced relative to each other along the $\log \epsilon$ axis to show some detail. The figure shows that there exists a region, $-3 \leq \log C_n(\epsilon) \leq 5$ approximately, where the curves are apparently close to being parallel. We take this to be the "scaling region" which presumably contains information on the attractor. Lower values of $\log C_n(\epsilon)$, which correspond to lower values of $\epsilon$ are affected by noise, and higher values, corresponding to higher values of $\epsilon$ begin to show "saturation" when $\epsilon$ becomes roughly of the size of the largest interpoint distances on the attractor.

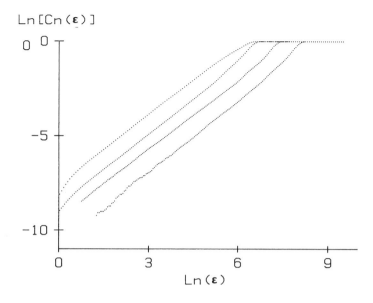

Figure 2. Log $C_n(\epsilon)$ vs. log $\epsilon$ for the Hénon map, $(x_{n+1}, y_{n+1}) = (y_n + 1 - 1.4x_n^2, 0.3x_n)$ for N points and embedding dimension = 4. From left to right, the graphs are for $N = 10^4$, $4 \times 10^3$, $1.2 \times 10^3$ and $0.5 \times 10^3$. The graphs have been displaced relative to each other along the log $\epsilon$ axis to show some detail.

To show this scaling property more clearly, we numerically differentiated each of the curves by doing least squares calculations of the slopes using seven-point segments on the curve. These slopes are plotted vs. log $C_n(\epsilon)$ in Figure 3. The top four graphs in the figure show individual plots of slopes vs log $C_n(\epsilon)$, the topmost being that for $10^4$ points, the number of points used decreasing from top to bottom. On the bottom graph, all four are superimposed on the same scale showing a "plateau" region where the curves coalesce and where the slopes are within 10% of each other. In this region the slopes are $1.24 \pm 0.02$ for N = $10^4$; $1.24 \pm 0.04$ for N = $4 \times 10^3$; $1.20 \pm 0.04$ for N = $1.2 \times 10^3$; and $1.28 \pm 0.09$ for N = 500. These results, together with the results of similar analysis of a variety of other systems which will be reported elsewhere[7] strongly suggest that it is possible to obtain some nontrivial dynamical information even from data sets as small as those which we used.

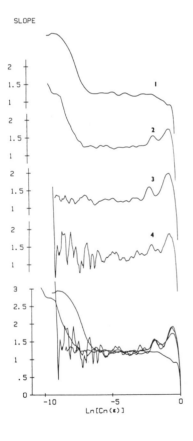

Figure 3. Slope vs log $C_n(\epsilon)$ for the curves shown in Fig. 2. The top four graphs, from top to bottom, are in decreasing order of N. The bottom graph shows all four superposed on each other.

Figure 4 shows Log $C_n(\epsilon)$ vs log $\epsilon$ for embedding dimensions 10 to 20 for a data set taken for particular laser operating conditions, Figure 5 shows the corresponding power spectrum wlhich displays the signature of a signal characterized by a single frequency and its harmonics, and Figure 6 shows the slope of Log $C_n(\epsilon)$ vs log $\epsilon$ plotted against log $C_n(\epsilon)$. Figure 6 clearly shows that for small values of log $C_n(\epsilon)$, the slopes increase with embedding dimension, as is expected for random noise. As log $C_n(\epsilon)$ increases, the different slopes coalesce to a plateau at $1.1 \pm 10\%$, before falling to zero at saturation.

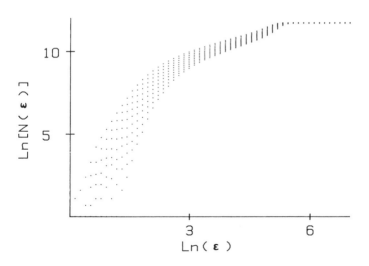

Figure 4.  Log $C_n(\epsilon)$ vs. log $\epsilon$ for a periodic (period 1) laser output intensity time series. The embedding dimensions are between 10 and 20.

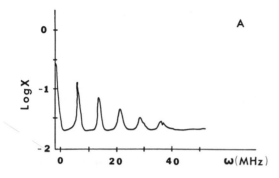

Figure 5. The power spectrum of the signal used to obtain Fig. 4, showing a fundamental frequency and its harmonics.

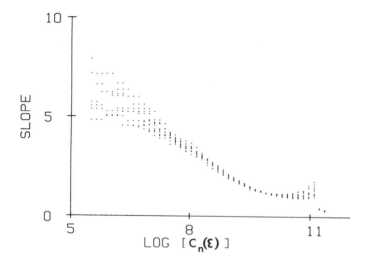

Figure 6. Slope vs Log $C_n(\epsilon)$ for embedding dimensions 10 – 20 for the signal used in Figs. 4 and 5. The slopes show a plateau at 1.1 ± 10%

Figures 7 to 9 are the corresponding curves for another data set. In this case, the spectrum suggests chaotic behavior which is confirmed by the existence of a plateau in the slopes plotted in Figure 9 at a value of 2.6 ± 10%.

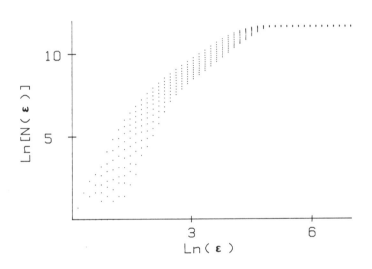

Figure 7. Log $C_n(\epsilon)$ vs. log$\epsilon$ for a chaotic laser output. Embedding dimensions 10 – 20.

Figure 10 shows the principal pulsing frequencies of the laser as a function of the voltage across the piezoelectric crystal ($V_{PZT}$), which is a measure of the cavity detuning. The x's indicate where the peaks were broad or imposed on a broadband background presumably indicating chaotic behavior as in Fig 8. Figure 11 is a plot of of $D_2$ and $K_2$ vs. $V_{PZT}$. It is seen that there is at least qualitative agreement among the results shown

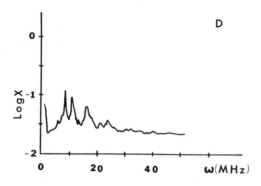

Figure 8. The power spectrum of the signal used to obtain Fig. 7, showing peaks superimposed on a broad background.

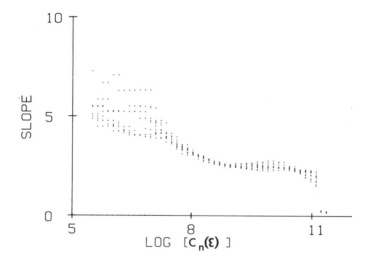

Figure 9.  Slope vs Log $C_n(\epsilon)$ for embedding dimensions 10 – 20 for the signal used in Figs. 7 and 8. The slopes show a plateau at 2.6 ± 10%

in Figures 4 to 9.  Values of $V_{PZT}$ which gave spectra characteristic of a single pulsing frequency or of two commensurate frequencies yield dimensions close to 1 while those values which gave apparently chaotic

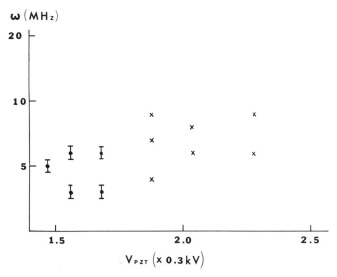

Figure 10. Principal pulsing frequencies vs. voltage across the piezoelectric transducer, $V_{PZT}$. The x's denote locations of broad peaks, or of peaks superposed on a broad background.

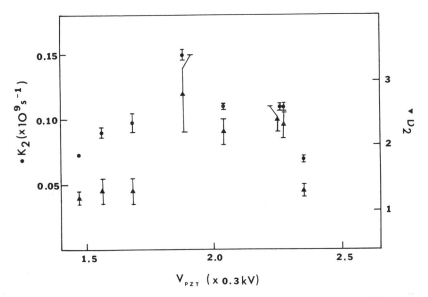

Figure 11. Order-2 information dimension, $D_2$, and order-2 Kolmogoroff entropy, $K_2$, vs. voltage across the piezoelectric transducer, $V_{PZT}$.

spectra yielded dimensions between 2 and 3. The changes in the values of $K_2$ also suggest changes in the attractor that are qualitatively similar to those inferred from the changes in $D_2$.

## Conclusions

The foregoing results show that the broadband pulsing exhibited by a single mode, inhomogeneously-broadened, single-mode ring laser can be described in terms of motion on a chaotic attractor of relatively low dimensionality. This agrees with similar conclusions drawn from a study of an analogous laser system in the Fabry-Perot configuration.[2]

These results also show that it is possible to obtain nontrivial dynamical information from time-sequence measurements of a single experimental variable in spite of limitations imposed by noise and relatively small data sets. They also demonstrate the robustness of the calculational techniques used; these techniques, used here to analyze data with characteristic frequencies in the megahertz range, had previously been used to analyze hydrodynamic, chemical and electrophysiological data with characteristic frequencies of a few hertz.[1]

## Acknowledgements

We have benefited greatly from discussions with colleagues J. P. Gollub, S. Cilliberto, A. Wolf, H. Swinney, J. D. Farmer, I. Procaccia, and J. R. Tredicce on the intricacies of calculating dimensions and entropies. This work was supported in part by a grant from the National Science Foundation, ECS82-10263.

# References

1.  A. Brandstater, J. Swift, H. Swinney, A. Wolf, D. Farmer, E. Jen, and J. Gutchfield, Phys. Rev. Lett. **51**, 1442 (1983); A. Brandstater and H. L. Swinney, in *Fluctuations and Sensitivity in Nonequilibrium Systems*, W. Horsthemke and D. K. Kondepudi, eds., Vol. 1 of Proceedings in Physics (Springer-Verlag, Berlin, 1984), p. 166; S. Ciliberto and J. P. Gollub, J. Fluid. Mech. (to be published); P. A. Rapp, I. D. Zimmerman, A. M. Albano, G. C. de Guzman, and N. N. Greenbaum, Phys. Lett. **110A**, 335 (1985).

2.  A. M. Albano, J. Abounadi, T. H. Chyba, C. E. Searle, S. Yong, R. S. Gioggia, and N. B. Abraham, J. Opt. Soc. Am. B **2**, 47 (1985).

3.  P. Grassberger and I. Procaccia, Phys. Rev. A **28**, 2591 (1983); Phys. Rev. Lett. **50**, 346 (1983); Physica **13D**, 34 (1984); A. Ben Mizrachi, I. Procaccia, and P. Grassberger, Phys. Rev. A **29**, 975 (1984); A. Cohen and I. Procaccia, Phys. Rev. A **31**, 1872 (1985).

4.  L. M. Hoffer, T. H. Chyba, and N. B. Abraham, J. Opt. Soc. Am. B **2**, 102 (1985).

5.  L. E. Urbach, S.-N. Liu, and N. B. Abraham, in *Coherence and Quantum Optics V*, L. Mandel and E. Wolf, eds. (Plenum, New York, 1984), p. 593.

6.  N. B. Abraham, L. A. Lugiato, P. Mandel, L. M. Narducci, and D. K. Bandy, J. Opt. Soc. Am. B **2**, 35 (1985).

7.  N. B. Abraham, A. M. Albano, B. Das, G. de Guzman, S. Yong, R. S. Gioggia, G. P. Puccioni, and J. R. Tredicce, Phys. Lett. A (submitted).

## NEW RESEARCH DIRECTIONS FOR CHAOTIC PHENOMENA IN SOLID MECHANICS

Francis C. Moon

Department of Theoretical and Applied Mechanics

Cornell University, Ithaca, New York

United States of America

### ABSTRACT

A review of known chaotic phenomena in solid mechanics is given and suggestions for new research areas involving chaotic vibrations in deformable and rigid body mechanics is presented. These areas involve three dimensional dynamics of the elastica, cylindrical shells and nonlinear plates, nonlinear acoustic-structure interactions, rotary dynamics of rigid bodies and servo-control problems. In addition, structural dynamics problems with nonlinear constitutive properties such as plasticity and viscoelasticity remain an unexplored subject for the study of chaotic phenomena.

### 1.    Introduction

Chaotic oscillations have been observed in both thermo-fluid and solid mechanical systems. However, chaotic fluid phenomena have been portrayed as <u>fundamental</u> while chaotic dynamics of elastic or rigid body systems have often been characterized as analog experiments of specific mathematical equations or technical applications of the theory. In this article, I advance the thesis that chaotic dynamics of continuous solid and rigid body systems constitute important problems for the field of nonlinear physics in that they involve different types of nonlinearities than appear in the Navier-Stokes equations of fluid mechanics. I will also summarize studies performed to date in chaotic solid mechanics as well as suggest topics for new directions of study in nonlinear dynamics of solid continua.

Chaotic vibrations stem from nonlinearities in physical and

mathematical systems. In mechanical continua, nonlinear effects arise from a number of different sources which include the following:

a)  kinematics; e.g. nonlinear convective acceleration, Coriolis and centripetal accelerations

b)  constitutive relations; e.g. stress vs. strain or strain rate, wave speed vs. pressure

c)  boundary conditions; e.g. free surfaces in fluids, deformation dependent constraints

d)  nonlinear body forces; e.g. magnetic or electric fields

e)  geometric nonlinearities; associated with large deformations in structural solids such as beams, plates, shells.

In the classic Navier-Stokes equations of fluid mechanics, the nonlinearity resides primarily in the convective acceleration or kinematic terms. The viscous effects are almost always assumed to be linear. In solid continua, on the other hand, nonlinearities are more likely to reside in the stress-strain or strain rate constitutive relations, boundary conditions or geometric nonlinearities. For example, in metal plasticity the material may be elastic below some stress state but will flow once some yield condition is reached in the stress space. This bimodal constitutive behavior can lead to hysteresis and locked-in stresses. In spite of this significant nonlinear effect, no experiments to date have been conducted on possible chaotic phenomena in these systems.

In structural mechanics of solids involving beams, plates and shells, another type of nonlinearity can arise. This nonlinear effect is geometric in origin and appears in the terms involving balance of forces and moments when deformation and rotations of the plate or shell element become large.

Besides the obvious curiosity about the existence of regular versus strange attractors in solid mechanics, the question of predictability of time histories of deformation and motion in solid continua is of great importance to the computer simulation industry, especially when nonlinear constitutive properties exist and large deformations are involved. For example, recent studies of fractal basin boundaries

reflecting the parameter sensitivity of even nonchaotic motions of simple nonlinear systems have raised the question of the ability of computer codes to predict the outcome of complex nonlinear mechanical systems.

## 2. Chaotic Dynamics of Elastic Continua

The mechanics of a thin flexible rod, known as the elastica have
been studied for over a century. The static behavior of elastic rods or
beams have been well understood and the phenomenon of buckling under
axial compressive forces is common knowledge. However, nonlinear dynamics of the elastica have received little attention, especially for
vibrations which are three dimensional in character. In fact, almost no
work has been done on large amplitude motions of the elastica. It is
known that forced planar vibrations may become unstable for large enough
amplitude and will bifurcate to out of plane vibrations. But even for
the simplest of theoretical models, the mathematical analysis is formidable due to geometric nonlinearities.

Chaotic dynamics in a nonlinear beam have been observed (see next
section) under nonlinear body forces and nonlinear boundary conditions.[1-3] However, the effect of geometric nonlinearities on predictability of large dynamic motions of an elastic beam or rod have yet
to be fully explored. Recent numerical work by Maewal[4] on harmonically
forced vibrations of an elastica including geometric nonlinearities and
damping has revealed chaotic motions for small nonlinearities.

Similar statements can be made for the case of nonlinear plate and
shell dynamics. Dowell[5] has investigated flutter in a Von Karman plate
model due to fluid flow over the plate using numerical simulation but
experiments have yet to confirm these predictions. Irregular vibrations
of harmonically excited elastic cylindrical shells have been reported as
early as 1965[6-7] but in the recent decade of chaos research little or
no research on chaos in elastic shells seems to have been done.

Other problems in solid continua which should exhibit chaotic
dynamics but which have received little attention thus far include the
"fire hose instability" due to flow through on elastic tube (see e.g.
Sethna[8]) and the moving elastic threadline. High speed motion of thin
thread-like continua are an important technical problem in the textile

industry. (Prof. W.F. Ames of Georgia Tech made a laboratory film (University of Delaware, circa 1966) of very large oscillations of moving strings in which chaotic-like motions appeared.)

## 2.1    Planar Elastica with Multi-Equilibria

Chaotic dynamics of a planar elastica have been reported by the Author and coworkers.[1,9,10] To illustrate the nature of the nonlinearities in this problem we record the equation of motion as derived by Moon and Holmes.[1] The deformation of an element of the rod is characterized by displacements (u,v) and slope angle of the rod centerline $\theta$. When the axial deformations are neglected (inextensibility), constraint equations result between u, v, $\theta$, (see Figure 1),

$$(1+u')^2 + (v')^2 = 1, \ \tan \theta = u'/(1+u') , \tag{1}$$

where ( )' = $\partial/\partial s$ and s represents length along the deformed rod. The balance of momentum equations for an element of the rod take the form

$$M\ddot{v} = - \frac{\partial V}{\partial v} - G' \ ; \quad G = \left[ D\theta'' - \frac{\partial V}{\partial \theta} \right] (1+u') - Tv' ,$$

$$M\ddot{u} = - \frac{\partial V}{\partial u} + H' \ ; \quad H = \left[ D\theta'' - \frac{\partial V}{\partial \theta} \right] v' + T(1+u') . \tag{2}$$

$V$ represents a body force potential per unit length and T is the axial force in the rod. In this problem, there are no kinematic or constitutive nonlinearities. Also dissipative effects are not represented in (2). The nonlinear effects lie either in the force potential $V(u,v,\theta)$ or in the geometric terms of the form $\theta''u'$, $Tv'$ (T can be an unknown depending on end conditions in the rod).

In the work of Moon and Holmes[1] chaotic vibrations of a forced beam under nonlinear magnetic body forces were observed (Figure 1). Although the amplitudes were large (100 times the beam thickness, or 20% of the beam length) it is believed that the magnetic force nonlinearities in these experiments dominated the geometric nonlinearities.

A one mode truncation of the equations (1), (2) was performed for cantilever boundary conditions and the dynamics were modelled by a Duffing type ordinary differential equation for the modal amplitude of the beam A(t),

a)

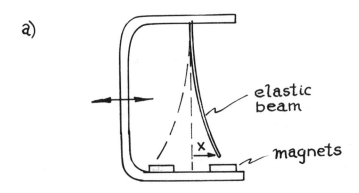

FIGURE 1
  a) Sketch of buckled elastic beam near two magnets.
  b) Poincaré map in the (A,Ȧ) plane for chaotic
     motions of the buckled beam.

b)

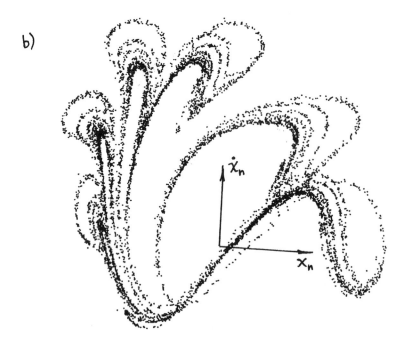

$$\ddot{A} + \gamma \dot{A} - \frac{1}{2} A(1-A^2) = f \cos \omega t. \tag{3}$$

Chaotic vibrations for this equation were first studied by Holmes[11] and predictions based on this model agreed well with low frequency experimental results for the elastic beam with magnetic forces. However, more recent work at Cornell and by others suggests that higher modes and geometric nonlinearities may be important for higher forcing levels or different frequencies and boundary conditions.

## Multi-Well Potential Problems

Several studies have been conducted at Cornell on chaotic vibrations of the elastic beam in a multi-well magnetic potential. Using rare earth permanent magnets placed near the tip of a steel cantilevered beam, two, three, and four static equilibrium problems have been investigated. In an earlier work the author showed that the static beam buckling problem with two magnets had the characteristics of a butterfly catastrophe.[1] In a later study (Golnaraghi and Moon[12]) the characteristics of forced chaotic motions in a four-well potential were explored beyond the critical forcing amplitude for chaos. A sample of this data is shown in Figure 2 which depicts the different types of dynamical motions as a function of forcing amplitude and frequency. This study showed that above the critical boundary for chaos, periodic islands of subharmonic motion could exist. Exploration at a finer scale revealed further structure suggesting the possibility of fractal boundaries.

Experimental Poincaré maps have played a crucial role in the investigation of chaos in the buckled beam. By synchronizing the measurement of the bending strain and strain rate with the harmonic forcing motion, two dimensional sections of the three dimensional attractor have revealed the fractal nature of the attractor. By varying the phase of the Poincaré section, one can map out cross sections around the entire attractor as shown in Figure 3. Because of the symmetry in the problem, the Poincaré section becomes inverted when the phase changed by 180°.

## Multiple Poincaré Sections

When the strange attractor lives in a phase space of one or more dimensions, the Poincaré section becomes three or more dimensional and a projection onto the plane does not reveal any fractal structure.

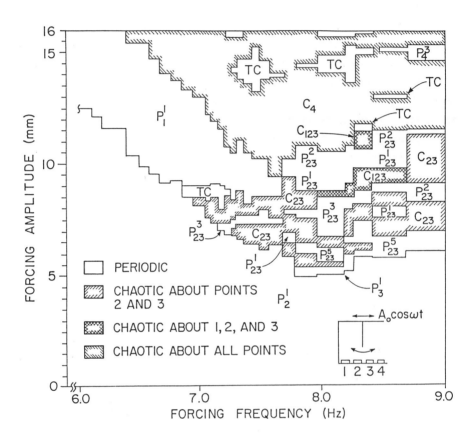

FIGURE 2
   Regions of chaotic and periodic vibrations in
   the forcing amplitude and frequency plane
   for the elastic beam in a four-well magnetic
   force potential.

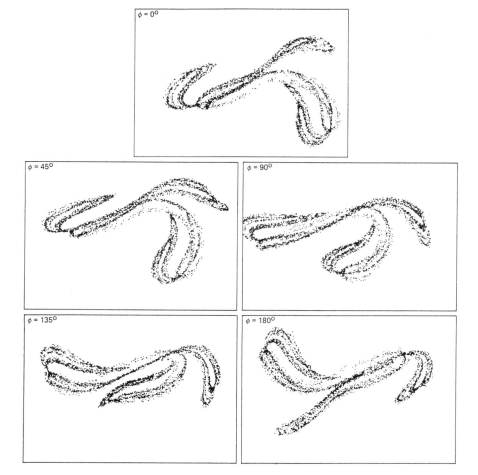

FIGURE 3
    Poincaré maps for chaotic motions of a buckled beam
    at different phases of the forcing amplitude.

However, we have recently developed a multiple Poincaré section technique for an attractor in a four dimensional phase space.[13]

An excursion beyond three dimensional phase space results when the buckled elastic beam is driven with <u>two</u> incommensurate driving frequencies. A one mode projection of this motion takes a form similar to (3),

$$\ddot{x} + \gamma\dot{x} + g(x) = f_1 \cos\phi_1 + f_2 \cos\phi_2 \, ,$$

$$\dot{\phi}_1 = \omega_1 \, , \tag{4}$$

$$\dot{\phi}_2 = \omega_2 \, .$$

In experiments on a buckled beam driven by two frequencies we have observed quasi-periodic vibrations with a transition to chaos.[13] In this technique a regular Poincaré section is used on the phase $\phi_1 = \omega_1 t$ which results in a three dimensional Poincaré map $(x_n, \dot{x}_n, \omega_2 t_n)$. To expose the fractal nature of this map we section the three dimensional set of points by choosing a window on the second phase, $\phi_2^* < \phi_2 < \phi_2^* + \Delta\phi_2$. Points falling in this window are projected on a plane. The results are shown in Figure 4 which exposes the fractal-like structure of the four dimensional attractor. A double Poincaré section has also been used by Lorenz in numerical studies[14], to expose the fractal nature of a strange attractor for a fourth order system of differential equations.

## 2.2   Chaotic Vibrations of a Three Dimensional Elastica

While the static behavior of elastic rods have been studied, especially as regards buckling phenomena,[15] little theoretical or experimental data exists on nonlinear vibrations of the elastica (Figure 5). The elastica is not only a model for a long thin structural column but also for elastic springs, coiled ropes, towed underwater cable or as a model for de-spooling fiber optics communications cable.

There are two types of nonlinearities in these problems--material based or stress-strain constitutive nonlinearities and geometric nonlinearities associated with large deformations. Large deformations are those displacements much greater than the small dimension of the rod or shell element. Large displacements also occur when there are significant changes in the curvature.

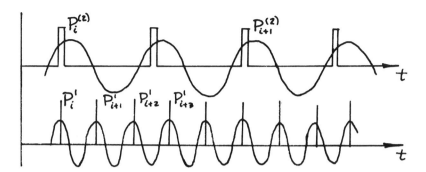

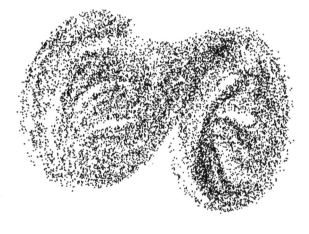

$P_i^1$ Map

$P_i^{(2)}$ Map

FIGURE 4

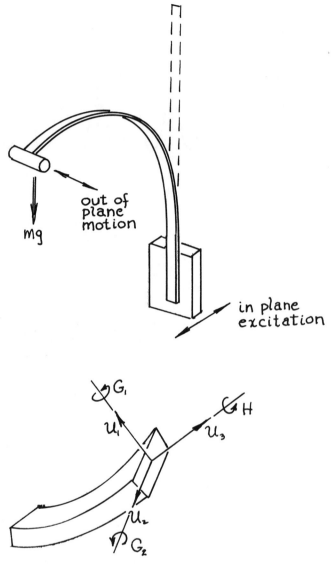

FIGURE 5
    Sketch of three dimensional motions of a
bent elastica.

The classical Kirchhoff-Love theory for the static behavior of the 3-D elastica may be found in Love (1892). In one sense the one dimensional elastic rod represents the simplest solid continuum. However, a glance at the equations required to describe the motions of this rod are convincing proof that understanding the dynamics of such a continuum is a formidable challenge.

In this theory the deformation is described in terms of three strain measures ($\kappa_1$, $\kappa_2$, $\tau$) which represent two bending curvatures and a twist or torsion measure. These curvature measures are in turn related in a nonlinear manner to the displacement ($u_1$, $u_2$, $u_3$) and angular position of the centerline or neutral axis of the elastica. The stress resultants at any cross section include three components of a vector couple ($G_1$, $G_2$, $H$) and three components of force ($N_1$, $N_2$, $T$). $G_1$, $G_2$ represent bending moments and $N_1$, $N_2$ represent shear forces in the cross section referenced to axis which move with the rod. $H$ is an axial torque, while $T$ represents an axial tension. In the simplest theory, the moments are linearly related to the curvatures ($G_1 = A\kappa_1$, $G_2 = B\kappa_2$, $H = C\tau$).

Without reproducing the complete theory we simply write a few of the equations to expose the nonlinearities. For example, the balance of linear momentum in the local $x_1$ direction takes the form

$$\frac{\partial N_1}{\partial s} - N_2 \tau + T\kappa_2 + f_1 = m \frac{\partial^2 u_1}{\partial t^2} . \tag{5}$$

The balance of angular momentum in the simplest theory, neglects the local rotary inertia. The $x_1$ component equation takes the form

$$A \frac{\partial \kappa_1}{\partial s} - (B - C) \kappa_2 \tau - N_2 = 0 . \tag{6}$$

In all, there are six linear momentum and angular momentum balance equations for the unknowns ($u_1$, $u_2$, $u_3$, $N_1$, $N_2$, $T$) as well as three nonlinear equations relating the curvatures ($\kappa_1$, $\kappa_2$, $\tau$) and the displacements ($u_1$, $u_2$, $u_3$). Few if any dynamic solutions of these equations are known.

In an earlier study by the author,[16] bending vibrations of an elastic rod of circular cross section showed out of plane chaotic dynamics when nonlinear magnetic body forces were present (see Figure 6). A small sphere of soft magnetic material was placed at the end of the rod and two permanent magnets were placed below the sphere which created a two-well potential. The periodically forced vibrations exhibited quasi-periodic as well as chaotic motions. A two-mode mathematical model of this system was proposed of the form

$$\ddot{x} + \delta\dot{x} + \frac{\partial V}{\partial x} = 0 \ ,$$
$$\ddot{y} + \eta\dot{y} + \frac{\partial V}{\partial y} = f \cos\omega t \ . \tag{7}$$

where $V(x,y)$ is a potential energy function for both the elastic restoring forces and the magnetic body forces. This model neglects the geometric nonlinearities and embodies only the nonlinear body forces on the sphere. An experimental criterion for chaos as a function of $(f,\omega)$ was found as shown in Figure 7. Theoretical criteria have been proposed for the planar motion of a beam in a two-well potential[3,11] based on equation (3), however, no criteria exists at present for the two dimensional motion of the type described by (7).

Recently we have begun further experiments on the elastica to assess the role of geometric nonlinearities in chaotic vibrations. One of the interesting phenomena we have seen in laboratory experiments is a bifurcation from planar periodic vibration to three dimensional chaotic looking vibrations at a large enough vibration force amplitude. We have also observed periodic sub-harmonic, quasi-periodic and chaotic vibrations. We believe the study of the coupling between modes (in this case in-plane and out of plane vibration modes) will uncover a rich spectrum of dynamic phenomena.

We are also investigating the usefulness of measuring the fractal dimension of this attractor in order to establish whether a finite mode model of the continuous elastica can be established.

### 2.3 Nonlinear Dynamics of Thin Cylindrical Shells

The understanding of dynamic shell elasticity is important for a variety of applications from rocket casings, underwater vehicles and even the manufacture of cylindrical cans for food processing. While much work has been done on the linear theory of shell vibrations, there

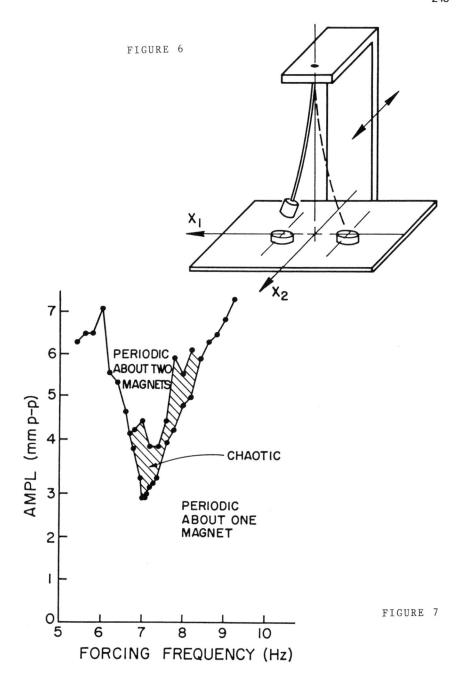

FIGURE 6

FIGURE 7

exists only a few experimental studies of nonlinear shell dynamics. In searching the literature, we have come across earlier work in the 1960's on nonlinear shell vibrations which suggests that chaotic vibrations of cylindrical shells can occur for sufficiently high forcing amplitude.

For example, Evensen[7] in a 1967 NASA technical note discussed the theory and some experiments on shell vibrations (Figure 8). Here he has studied the problem of nonlinear mode coupling between two cylindrical mode frequencies. If $w(x,\theta,t)$ represents the radial shell displacement he represents this displacement in the form

$$
\begin{aligned}
w = [A(t) \cos n\theta + B(t) \sin n\theta]\sin \frac{m\pi x}{L} \\
+ \frac{n^2}{4R} [A^2(t) + B^2(t)]\sin^2 \frac{m\pi x}{L} ,
\end{aligned}
\tag{8}
$$

where $(n,m)$ represent circumferential and axial wave numbers, R is the mean shell radius, and L is the length of the shell. He uses Galerkin's method to derive two coupled nonlinear equations for the two mode amplitudes $A(t)$, $B(t)$. In nondimensional form these become[7]

$$
\begin{aligned}
\ddot{A} + A = &-\frac{3}{8} \varepsilon A(A\ddot{A} + \dot{A}^2 + B\ddot{B} + \dot{B}^2) \\
&+ \varepsilon\tau A(A^2 + B^2) - \varepsilon^2\delta A(A^2 + B^2)^2 + Q \cos\omega t , \\
\ddot{B} + B = &-\frac{3}{8} \varepsilon B(B\ddot{B} + \dot{B}^2 + A\ddot{A} + \dot{A}^2) \\
&+ \varepsilon\tau B(B^2 + A^2) - \varepsilon^2\delta B(B^2 + A^2)^2 .
\end{aligned}
\tag{9}
$$

These equations possess unusual nonlinear terms when compared with the conventional systems studied in dynamical systems. They predict a softening type nonlinear response for periodic vibrations which are observed experimentally.[7] However, in both analog computer studies of the above equations (9) and in experiments, Evensen reported nonsteady vibrations for sufficiently large driving amplitudes. Thus this problem is an important candidate for the study of multi-mode quasi-periodic and chaotic vibrations. The cylindrical shell problem like the earlier spherical pendulum study by Miles[17] is an example of dynamics in a problem with symmetry. We are interested in how this symmetry is broken with small geometric perturbations.

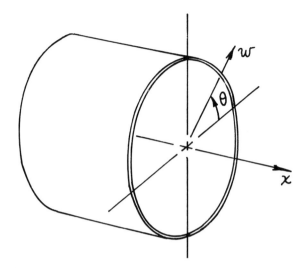

FIGURE 8
    Sketch of a cylindrical shell element.

FIGURE 9
    Four bilinear impact problems which
    exhibit chaotic dynamics.

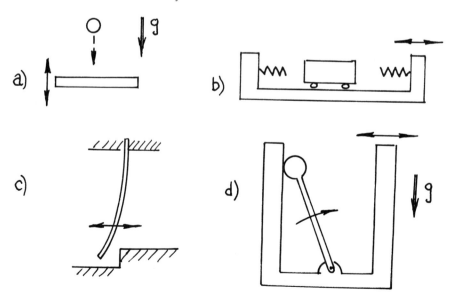

## 2.4   Impact Problems

A class of nonlinear problems which has received a great deal of attention is the impact of solids under periodic excitation (Figure 9). (In the physics literature, this is often called the Fermi accelerator problem which models the motion of electrons in electromagnetic fields, see e.g. Lichtenberg and Lieberman[18].) This class of problems is of interest to theoreticians since explicit analytical expressions for the Poincaré maps can be obtained by integrating the equation of motion between impacts. An energy loss impact law is assumed relating the relative normal velocities of the particle before and after impact. In one set of problems, the particle is acted on by gravity[18,19], and in another, the particle is restrained by a linear spring force between impacts (see e.g. Shaw[20,21]).

A variation of this problem was examined by Moon and Shaw[22] who looked at the forced motion of a cantilevered elastic beam in which the free end impacted a solid constraint when the amplitude of the end of the beam exceeded some value.

In all of these problems, chaotic motions were observed in both numerical simulation and experiments. These examples point the way to a class of problems in mechanical systems with many moving parts such as gears, shafts, cams, etc. All such systems involve gaps, play or backlash, and are subjected to impact forces. It is suspected that mechanical systems with backlash nonlinearities will exhibit chaotic vibrations in certain parameter regions of the problem.

In what might be the first commercial exploitation of chaotic dynamics, Hendricks, of IBM-Yorktowne Heights, has studied impact-induced chaos in print hammers[23]. He has demonstrated that one of the technical roadblocks to increasing the speed of impact printers is the fact that increased forcing frequency leads to chaotic print hammer response. He has performed experiments using Poincaré maps to demonstrate the existence of strange attractor behavior and has also developed a technically realistic mathematical model to simulate the nonlinear dynamics. He has proposed feedback control to increase print hammer frequency and quench chaotic behavior.

## 2.5    Chaotic Acoustic Phenomena

One of the goals of our current research program is to explore bifurcations and chaotic behavior in higher dimensional phase spaces.  In
the acoustics problem, we are examining a one dimensional acoustic system with nonlinear boundary or impedance conditions.  The problem represents a coupling between a linear wave equation and a nonlinear ordinary
differential equation which embodies the dynamics of the boundary impedance.  We have already obtained experimental results in a preliminary
experiment using an air filled tube coupled to a spring loaded end plug
(Figures 10, 11).  These results show that a high frequency acoustic
mode can drive a low frequency nonlinear mechanical oscillator into
chaotic vibration.  Chaotic modulation of the acoustic signal as measured with a microphone near the end of the tube is also observed (see
Figure 11).

We are studying the correlation in phase between sound pressure
levels along the tube in order to establish the existence of chaotic
waves.  Also of interest is to determine whether the chaotic dynamics of
this continuum can be modeled by a finite set of modes.  To this end, we
are attempting to measure the fractal dimension of the attractor.

## 3.    Chaos in Rigid Body Dynamics and Control Systems

The study of chaos in rigid body dynamics has not received much
attention.  The subject of rigid body dynamics is taught in elementary
physics and mechanics, yet the nonlinear nature of the equations of
motion preclude application to all but the simplest problems such as the
pendulum or moment free body.  The sources of the nonlinearities lie in
the kinematics as well as the force and moment equations.  If $\underline{V}_C$ represents the velocity of the center of mass, and $\underline{\omega}$ represents the rotation
rate vector of the body, the Newton-Euler equations of motion take the
form

$$M\dot{\underline{V}}_C = \underline{F} ,$$
$$I_1\dot{\omega}_1 + (I_3 - I_2) \omega_2\omega_3 = G_1 ,$$
$$I_2\dot{\omega}_2 + (I_1 - I_3) \omega_1\omega_3 = G_2 ,$$
$$I_3\dot{\omega}_3 + (I_2 - I_1) \omega_1\omega_2 = G_3 ,$$

(10)

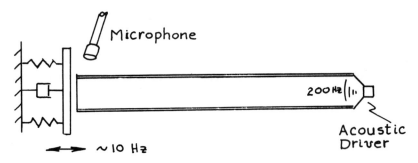

FIGURE 10
    Sketch of experiment showing the coupling
    between a one dimensional acoustic system
    and a one degree of freedom elastic end plug.

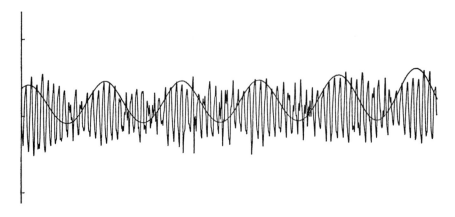

FIGURE 11
    Chaotic modulation of sound field due to
    nonlinear acoustic-structure interaction.

where the force and moment vectors, $\underset{\sim}{F}$, $\underset{\sim}{G}$, might be functions of general-ized coordinates and linear and angular velocities. The nonlinear acceleration terms, $\omega_2\omega_3$, etc. are similar to those in the Lorenz equa-tions.

One example of a study of chaotic dynamics of Euler's equation was published by Leipnik and Newton[24]. They looked at the case where the applied moment is proportional to the rotation vector, i.e.

$$\underset{\sim}{G} = \underset{\approx}{A} \cdot \underset{\sim}{\omega}$$

where $\underset{\approx}{A}$ is a torque feedback matrix. They reported observing a "double strange attractor," so-called for the shape the attractor takes in the $(\omega_1, \omega_2, \omega_3)$ phase space. They used both analog and digital simulation.

In the next section, we briefly describe three research problems we are conducting at Cornell involving rigid body dynamics and feedback control.

### 3.1  Nonlinear Mechanical Position Controller with Linear Feedback

In a recently completed study in our laboratory, we observed cha-otic motions of a nonlinear mechanical system with a linear feedback servo positioning motor (Figure 12). The equations of this system are of the form

$$\ddot{x} + g(\dot{x}) + h(x) = -z \ ,$$
$$\dot{z} + \alpha z = \Gamma_1(x-f(t)) + \Gamma_2\dot{x} \ , \qquad (11)$$
$$f(t+\tau) = f(t) \ .$$

Both g( ) and h( ) can be nonlinear. This is a dynamical system in a four dimensional phase space. We have explored chaos as a function of the control gains $(\Gamma_1,\Gamma_2)$. Recently Wiggins, in a doctoral dissertation at Cornell, has derived a theory using the Melnikov method to analyze a feedback control problem similar to (11) with a three dimensional Poincaré map space. We have hopes that these techniques may be useful for multi-degree of freedom problems. Holmes[25] has also published theoretical results for a system similar to (11).

An earlier study by Moon and Holmes[26] reported chaotic motions in an autonomous mechanical system with feedback control in which the

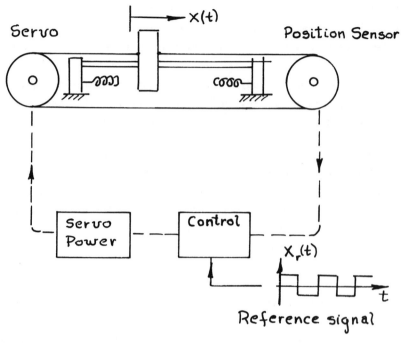

FIGURE 12
   Sketch of a one degree of freedom mechanical
   positioning device with servo motor feedback.

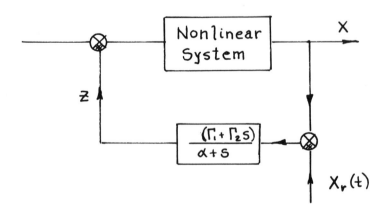

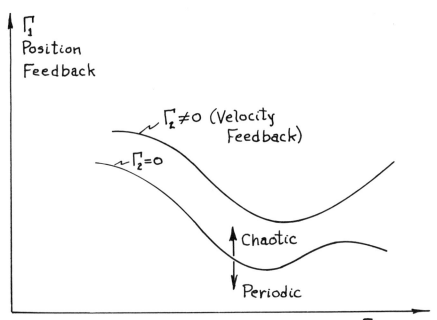

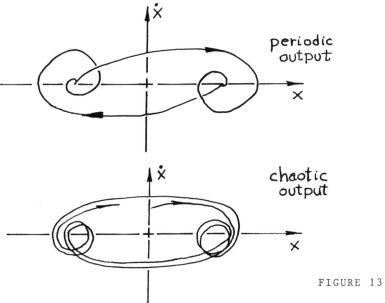

FIGURE 13

equilibrium position first loses stability by a Hopf bifurcation, and then undergoes chaotic dynamics as gain is increased.

In the experimental study to be published by Golnaraghi and the author[27], the boundary between chaotic and periodic motions under periodic control forces was explored. The nonlinear forces were elastic mechanical stops which constrained the overshoot dynamics of the moving mass. A sketch of the qualitative results is shown in Figure 13. For fixed gains, an increase in control force frequency brought about chaotic vibrations. However, using rate feedback, one could return the system to periodic behavior.

### 3.2 Forced Oscillations of a Rigid Magnetic Rotor

Chaotic motions of a forced pendulum or rotor have received a great deal of attention because it satisfies the same equation as a Josephson junction. Experiments on the dynamics of a magnetized rotor in oscillating magnetic fields were reported by Croquette and Poitou[28]. In their experiments they looked at the zero damping or Hamiltonian limit. Recently, at Cornell, experiments have been performed on a rotor with restoring torque and applied periodic torques[29]. The mathematical model represents a dipole in crossed steady and oscillating magnetic fields. The nondimensionalized equation of motion is given by

$$\ddot{\theta} + \delta\dot{\theta} + \sin\theta = f \cos\theta \cos\omega t. \tag{12}$$

The experiments were performed in the moderate damping limit ($\delta=0.5$) and Poincaré maps of the motion demonstrated the existence of a strange attractor for sufficiently high damping (Figure 14). The method of Melnikov was used to derive a condition for the existence of homoclinic orbits in the Poincaré map.

$$f > \frac{4\delta}{\pi\omega^2} \cosh\left(\frac{\pi\omega}{2}\right). \tag{13}$$

The appearance of homoclinic orbits implies the existence of horseshoe type maps which are thought to be a necessary precursor to chaos. The criterion (13) seems to provide a good lower bound for the experimentally determined values of $(f,\omega)$ which lead to chaotic motions. The use of the Melnikov method to derive a lower bound on the criterion for chaos has also been successfully used in the example of the buckled

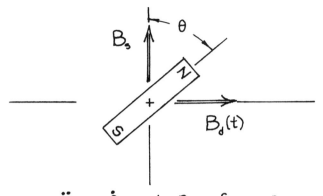

$$\ddot{\theta} + \gamma\dot{\theta} + \sin\theta = f\cos\theta\cos\omega t$$

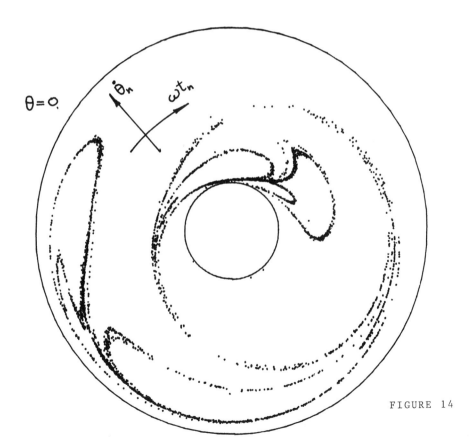

FIGURE 14

beam. However, the extension of this technique to derive a criterion for chaotic motion for two or more degrees of freedom systems remains a challenge for theorists.

### 3.3 Chaotic Dynamics of Serial Links with Multiple Feedback Loops

With a view toward robotic applications, we are currently investigating chaotic motions of a serial linked arm under periodic commands as a function of multiple feedback loops that control the relative motion of each link in the arm. The goal here is to investigate how chaotic vibrations can arise in such systems and how feedback can produce or quench such parameter sensitive dynamics. A two link mechanism has equations of the form,

$$D_{111}\ \ddot{\theta}_1 + D_{112}\dot{\theta}_1\dot{\theta}_2 + k_1\theta_1 + c_1\dot{\theta}_1 = T_1(z)\ ,$$

$$D_{222}\ \ddot{\theta}_2 + D_{211}\dot{\theta}_1 + k_2\theta_2 + c_2\dot{\theta}_2 = T_2(t)\ , \tag{14}$$

where

$$D_{ijk} = g(\theta_1,\theta_2)\ .$$

The nonlinearities in this system represent centripetal ($\dot{\theta}_1^2$) and Coriolis accelerations ($\dot{\theta}_1\dot{\theta}_2$). The linear terms in $\theta_1,\dot{\theta}_1$, etc. can represent either stiffness and damping or proportional and rate feedback terms. While such nonlinear terms are common in equations of robotic devices, almost no reports of the effect of such nonlinearities on the dynamics of robotic devices have appeared. The third order system studied by Leipnik and Newton[24] on the spinning rigid body contains similar product nonlinearities but the robotic system is of fourth order or higher. As robotic devices became lighter in weight and faster in cycle frequency, questions concerning flexibility of links, control and nonlinear dynamics and chaos will become more important.

### 4. Research Directions for Chaos in Solid Mechanics

One of the basic questions in nonlinear mechanics of solid systems is: How do various types of nonlinearities affect predictability and the occurrence of chaotic phenomena? In rigid systems, only the effects of nonlinear restraining forces and limited control forces algorithms have been explored. In deformable solids, only nonlinear body forces and

nonlinear boundary conditions have received significant attention and
recently some work on geometric nonlinearities. An agenda for new
research in this area should include the following:

i)     the study of predictability and chaos in solids with constitutive nonlinearities such as plasticity and creep;

ii)    experimental and theoretical analysis of geometric nonlinearities in large deformations of elastic rods, plates
and shells;

iii)   the extension of mathematical methods in nonlinear dynamics
for low order system to multi-degree of freedom problems--
especially multi-mode interactions;

iv)   development of mathematical techniques in nonlinear dynamics
for direct application to partial differential equations of
nonlinear continuum mechanics;

v)     develop criteria to determine when fractal basin boundaries
will appear in nonlinear systems with many degrees of freedom;

vi)    determine the statistical nature of chaotic vibrations in
solid systems and the effects of outside noise;

vii)   investigate possible connections between chaotic dynamics
and microphysical failure mechanisms such as fracture,
fatigue, friction and wear.

Finally, from an engineering perspective, one should ask:

a)     how can one design to avoid chaotic motions;

b)     how can one use control to suppress chaotic motions when
they occur; and if all else fails,

c)     how does one live with chaotic phenomena and unpredictability in technical devices.

## REFERENCES

1.  Moon, F.C. and Holmes, P.J., "A magnetoelastic strange attractor," J. Sound and Vibration 65(2) (1979), 279-296.

2.  Holmes, P.J. and Marsden, J.E., "A partial differential equation with infinitely many period orbits: chaotic oscillations of a forced beam," Archive for Rational Mechanics and Analysis 76 (1981), 135-166.

3.  Moon, F.C., "Experiments on chaotic motions of a forced nonlinear oscillator: strange attractors," J. Applied Mechanics 47(3) (1980), 638-644.

4.  Maewal, A., "Chaos in a harmonically excited elastic beam," to appear in J. Applied Mechanics.

5.  Dowell, E.H., "Flutter of a buckled plate as an example of chaotic motion of a deterministic autonomous system," J. Sound and Vibrations 85(3) (1982), 333-344.

6.  Olson, M.D., "Some experimental observations on nonlinear vibration of cylindrical shells," AIAA J. 3 (1965), 1775-1777.

7.  Evenson, D.A., Nonlinear Flexural Vibrations of Thin-Walled Circular Cylinders, NASA Technical Note, NASA TN D-4090, August 1967.

8.  Bajaj, A.K. and Sethna, P.R., "Bifurcation in three-dimensional motions of articulated tubes," J. Applied Mechanics 49 (1982) Part I (606-611), Part II (612-618).

9.  Moon, F.C., "Fractal boundary for chaos in a two-state mechanical oscillator," Physical Review Letters 53(10) (1984), 962-964.

10. Moon, F.C. and Li, G.-X., "The fractal dimension of the two-well potential strange attractor," Physica 17D (1985), 99-108.

11. Holmes, P.J., "A nonlinear oscillator with a strange attractor," Phil. Trans. Royal Soc. Lond. 292, No. 1394, (1979), 419-448.

12. Golnaraghi, M. and Moon, F.C., Chaotic Vibrations of an Elastic Beam in a Four-Well Potential Force Field, Cornell University Report, Department of Theoretical and Applied Mechanics (1984).

13. Moon, F.C. and Holmes, W.T., "Double Poincaré sections of a quasi-periodically forced chaotic attractor," Physics Letters 111A(4) (1985), 157-160.

14. Lorenz, E.N., Physica 13D (1984) 90-114.

15. Antman, S.S. and Kenney, C.S., "Large buckled states of nonlinearly elastic rods under torsion, thrust and gravity," Arch. Rational Mech. Anal. 76 (1981), 2889-338.

16. Moon, F.C., "Experimental models for strange attractor vibrations in elastic systems," New Approaches to Nonlinear Problems in Dynamics, P.J. Holmes, ed., SIAM (1980), 487-495.

17. Miles, J.W., "Stability of forced oscillations of a spherical pendulum," Quart. Applied Math XX(1) (1962), 21-32.

18. Lichtenberg, A.J. and Lieberman, M.A. Regular and Stochastic Motion, Springer-Verlag, 1983.

19. Holmes, P.J., "The dynamics of repeated impacts with a sinusoidally vibrating table," J. of Sound and Vibration 84 (1982), 173-189.

20. Shaw, S.W. and Holmes, P.J., "A periodically forced piecewise linear oscillator," J. of Sound and Vibration 90 (1983), 129-155.

21. Shaw, S.W., "The dynamics of a harmonically excited system having rigid amplitude constraints: Parts 1, 2," J. Applied Mechanics 52 (1985).

22. Moon, F.C. and S.W. Shaw, "Chaotic vibrations of a beam with non-linear boundary conditions," Int. J. Non-Linear Mechanics 18(16) (1983), 465-477.

23. Hendricks, F., "Bounce and chaotic motion in impact print hammers," IBM J. Research and Development 27(3) (1983), 273-280.

24. Leipnik, R.B. and Newton, T.A., "Double strange attractors in rigid body motion with linear feedback control," Physics Letters 86A(2) (1981), 63-67.

25. Holmes, P.J., "Dynamics of a nonlinear oscillator with feedback control: Part 1, 2," J. Dynamic Systems, Measurement and Control 107 (1985), 159-165.

26. Holmes, P.J. and Moon, F.C., "Strange attractors and chaos in
     Nonlinear Mechanics," J. Applied Mechanics 50 (1983),
     1021-1032.

27. Golnaraghi, M. and Moon, F.C., The Effect of Linear Feedback on the
     Chaotic Motions of a Mechanical Servo Positioner, Cornell
     University Report, Department of Theoretical and Applied
     Mechanics (1985).

28. Croquette, V. and Poitou, C., J. Phys. Letters 42 (1981), L-537.

29. Moon, F.C., Cusumano, J., and Holmes, P.J., Evidence for Homoclinic
     Orbits as a Precursor to Chaos in a Magnetic Pendulum, Cornell
     University Report, Department of Theoretical and Applied
     Mechanics (1985).

# PRIME TIMES: THE DISTRIBUTION OF SINGULARITIES
# IN HYDROPHOBIC FREE ENERGY OF PROTEINS

Arnold J. Mandell

Laboratory of Biological Dynamics and Theoretical Medicine

University of California, San Diego (M-003)

La Jolla, California 92093

The following remarks touch on the many still inadequately posed and open questions in our efforts to understand the relationships between amino acid sequence-generated secondary structures and the mechanisms of action of enzyme, membrane receptor, and transport proteins. Of particular interest are allosteric proteins that undergo functionally significant conformational transition in response to relatively small changes in their environments. The enzymes aspartate transcarbamylase and tyrosine hydroxylase, the membrane ion conductance-mediating nicotinic cholinergic receptor, and the oxygen-binding transport macromolecules myoglobin and hemoglobin are good examples. The relevance of a global dynamical approach to their functions is suggested by x-ray crystallographic evidence that their regulatory and work sites may be as far apart as 30 Å.

A pervasive and hidden thermodynamic symmetry, enthalpy-entropy compensation, dominates the hydrophobic free energy of protein function. For example, neither mass nor density influences retention on hydrophobic liquid partition columns. Thus, the dynamical problem is reduced to one involving a two-dimensional enthalpy-entropy phase space: sets of orbits distributed like geodesics on a surface of negative curvature which we view as hyperbolic Farey-Anosov maps of the plane. Their trigonometric transformation yields hydrophobic modes, $z \in C$, which, like the singularities of the hydrophobic free energy found

Written while the author was a Visiting Fellow, Institut des Hautes Etudes Scientifiques, Bures-sur-Yvette, France. The analogue and digital simulations and calculations of amino acid hydrophobic free energy sequence statistics were supported by ARO DAAG20-83-K-0069 and DOD DAAG29-84-G-0072.

in proteins, scale time as complex, low order primes, $\Phi n \in \mathbf{C}$. These modes, $\Phi z$, can be recovered as universal singularities in the power spectra of critically nonlinear $(r = r_c)$ two-parameter $(r, \lambda)$ hydrophobic free energy maps and cross sections of their flows on the plane.

A universal exponent, $\alpha = 0.618...$ or $2/3$, derived from thermodynamic and dimensional considerations, describes the scaling of the orbit lengths of the geodesics, $\Phi z$, and gives $\alpha^{-1}/\alpha = (3/2)/(2/3) \sim \exp^{-2.27}$ as the universal power law of the protein's enthalpy-entropy distribution. A measure on the regions of loss of geodesics in parameter space $(r, \lambda)$ as a function of orbit length $\Phi z$ yields a power law of $\exp^{-2.28}$ when normalized by the Euler prime function $\Phi n$ in generic maps of a circle at $r = r_c$. The density of primes serves as a limit on the accumulation of geodesic orbits as a function of the parameter space partition modulated by the nonlinear coupling term $r$ on $\mathbf{R}^2$.

Specific signal-sensitive, reversible tangent bifurcations from the universal spectrum at $r = r_c$ yield bursting behavior that overcomes short range repulsion via intermittent increases in intramacromolecular pressure as well as linear-in-t diffusivity via turbulent mixing; this low energy requiring transition ameliorates rate-limiting influences on the binding, catalysis, and transport work of proteins.

The hyperbolic dynamics of the hydrophobic free energy of amino acid side chains in proteins serve as the mechanism of generation, storage, and delivery of the small, surface tension-like hydrophobic free energy of the macromolecule-solvent interface which proteins use for their functions.

I. Enthalpy-Entropy Compensation: The Thermodynamic Symmetry of Proteins

Polypeptide chains of amino acids minimize their contact with the flickering clusters of strained hydrogen-bonded cages of water[1], rolling up into a state of reduced chain entropy along with a compensatory increase in enthalpy, measurable as a change in heat capacity[2]. The hydrophobic free energy of the protein-solvent system, $u$, is a function of this enthalpy, $H(u)$, and its temperature-dependent entropy, $S(u)$. $F(u)$ can be envisioned as a narrowly bounded, enthalpy-entropy compensated, surface tension-like force at the macromolecule-solvent interface[3] such that

$$\Delta H(u) = \Delta H_0(u) + \alpha \Delta S(u) \tag{1}$$

in kcal/mol, with $\alpha$ the proportionality constant. From the second law,

$$\Delta F(u) = \Delta H(u) - T^\circ S(u) . \tag{2}$$

We note that $\alpha$ serves as an isokinetic temperature for the entropy $S(u)$, which we treat as complex: $S(u) \in C$. From (1) and (2) ,

$$\Delta F(u) = \Delta H_0(u) - (T^\circ - \alpha)\Delta S(u) . \tag{3}$$

It might be concluded that, because $\alpha$ varies with $T^\circ$, $\Delta F(u) = \Delta H_0(u)$, and a generalized rate constant for processes involving amino acid side chain-derived hydrophobic free energy, $k_i = k_0 e^{-F(u)/RT}$ would always be the same for a particular protein, $k_i = k$. However, we know that is not the case. For example, the allosteric enzymes aspartate transcarbamylase[4,5] and tyrosine hydroxylase[6,7] demonstrate Devil's staircases[8] of $k_i$. For membrane protein binding of ligands, such multiple saturation plateaus are characteristic[9], and the low energy transition (3.7 kcal/mol) between two oxygen binding states of hemoglobin, $k_i \rightarrow k_j$, serves as the best studied example[10]. Combining (1) and (2), in light of the multiplicity of states,

$$\Delta F(u) = (T^\circ/\alpha)\Delta H_0(u) + (1 - T^\circ/\alpha)\Delta H(u) \tag{4}$$

yields the true identity of $\alpha$: the scaling on a complex temperature-related entropy-compensated enthalpy, a critical exponent of a system whose work consists of multiple phase transitions in $\Delta^2 S(u)$. The transconformations $k_1 \leftrightarrow k_2 \leftrightarrow k_3...$ are responsible for the maintenance of relationship (1). We note that renormalization of the parameter distance between $k_i$ and $k_{i+1}$ occurs at changing levels of allosteric ligands[7], which when taken together with a Devil's staircase of second-order phase transitions in $k_i$, suggests the presence of a quasi-periodic, nonlinear system at critical coupling. Experimental findings[2,10] suggest that the work of allosteric proteins involves large changes in the enthalpy-compensated entropy for small net changes in the free energy. That the complex entropy scaling exponent $\alpha$ may be fractional is suggested by the wide range of relaxation times that have been determined in physical studies of proteins[7]: $C(\tau) \sim 10^{-12} \cdots 10^{4+}$. We can visualize hydrophobic aggregation as generating a fractal structure a bit like emerging globules in a salted soap solution.

In the context of a finite moment, equilibrium statistical mechanics of $u$, where $u_1$, $u_2$ ...$u_n$ represent its equivalent states; $\sum_i p_i u_i$, the average value of the functions; $u(i) = u_i$; and $g(u) = -\log u$ is continuous on $[0,1]$; the entropy of $u$ can be represented in abstract logarithmic units of information as

$$S(p_1, p_2 \cdots p_n) = \sum_{i=1}^{n} - p_i \log p_i \tag{5a}$$

$$S(u)(p_1, p_2 \cdots p_n) = \sum_{i=1}^{n} - p_i \log p_i + \sum_{i=1}^{n} p_i u_i . \tag{5b}$$

Because the distribution function of states of $u$, $P(u)$, has the strongly attracting basin of a central limit

$$\frac{u^2 \int d P(u)}{\int\limits_{u \to 0} u^2 P(u)} \to 0 , \tag{6}$$

an invariant measure $\mu$ on the entropy of this equilibrium Gibbs state can be defined[11] as

$$\mu[S(u)] = P(u_j) = \frac{e^{u_j}}{\sum\limits_{i}^{n} e^{u_i}} . \tag{7}$$

We complexify (7) as a characteristic function of the distribution $P(u)$ with expectation $\delta$, weight $\gamma$, distributional skew $\beta$, and characteristic exponent $\alpha$ describing the complex entropy as a measure on the spectral convergence, and we put the transformation into similar abstract logarithmic units of information[12] as

$$\log P_u(z) = i \delta z - \gamma |z|^\alpha (1 + i \beta z) . \tag{8}$$

When $\beta = 0$ and $\alpha = 2$, (7) and (8) are conjugate.

In contrast to this classical thermodynamic picture of a Gibbs ensemble of equivalent $u$'s, the Devil's staircase of concentric hierarchical hydrophobic free energy modes of protein dynamics consists of basins of weak and partial attraction, fractionally scaling $\alpha$-helical spirals in the neighborhoods of multiple hyperbolic equilibria. Minimizing $F(u)$ by reducing the surface of contact between amino acid side chains and solvent, with which $F(u)$ is well correlated[13,14], the globular protein assumes a spherical shape. With $l$ as its characteristic length scale,

$$\text{mass (Daltons)} \sim \text{volume} \sim l^{3} \tag{9a}$$

$$\text{surface} \sim l^{2} \tag{9b}$$

$$F(u) \sim \text{mass}^{2/3} \tag{9c}$$

and the hydrophobic coherent mode scaling exponent $\alpha \sim 2/3$. This estimate appears to be valid up to 50,000 Daltons, the average size of a protein monomer, as determined by x-ray crystallography[15]. The finding suggests at least a homeomorphic continuity in the transition from the primary sequence to the tertiary structure. Its use later as the mode scaling ratio in the Fourier transformation of sequences of amino acid hydrophobic free energies and the parameter of a Weierstrass-like cosine series describing their Farey-Anosov maps and flows necessitates the definition of $\alpha$ in this hierarchical context as an invariant operator on a multiplicative group of reals in the space of complex vectors.

Another estimate of the geometrically proportional operator $\alpha \in [0,1]$ derives from the limit on the ratio of emergent hydrophobic modes, $z_i$.

$$\frac{z_i}{z_{i+1}}, \frac{z_{i+1}}{z_i + z_{i+1}} \ldots \tag{10a}$$

$$\frac{z_i}{z_{i+1}} = \alpha \tag{10b}$$

$$\lim \alpha = \left| \frac{1}{\alpha + 1} - \epsilon \right| \leqslant |\alpha - \epsilon| \sim \frac{\sqrt{5} - 1}{2} = 0.618\ldots \tag{10c}$$

Since the dynamics of $u$ are dependent on $\alpha$ and not its bases, we normalize each of the 20 amino acids to a hypothetical unit mass, then use estimates on the hydrophobic free energy of individual amino acids[16] (Table 1). With Fourier transformation of these hydrophobic free energy sequences in over 100 polypeptides and proteins, we found the commonest modes to be of residue wavelength $\sim$ 2, 3, 5, 7-8, 11, and 13. This series of low-order prime numbers generates adjacent mode ratios that have an average value of $\alpha \sim 2/3$.

Expanding $|F(u)|^{\alpha}$ in the complex domain, using (10c) as a hydrophobic free energy mode aggregation operator,

$$F_u(z) = [\Gamma(\alpha + 1)]^{-1} \int_0^{\infty} z(u) e^{-(\alpha + 1)} \, dz(u) \tag{11a}$$

where

$$\Gamma(\alpha + 1) = \int_0^\infty [F(u)]^\alpha e^{-u} \, du \ . \tag{11b}$$

This generates a convolutionally stable divergent trigonometric series like (8) when $1 < \alpha < 2$.

| Table 1 | | | |
|---|---|---|---|
| Hydrophobic Free Energies Ethanol-Water Partition $(F(u) = \text{kcal/mol})$ | | | |
| Ala | 0.87 | Leu | 2.17 |
| Arg | 0.85 | Lys | 1.64 |
| Asn | 0.09 | Met | 1.67 |
| Asp | 0.66 | Phe | 2.87 |
| Cys | 1.52 | Pro | 2.77 |
| Glu | 0.67 | Ser | 0.07 |
| Gln | 0.00 | Thr | 0.07 |
| Gly | 0.10 | Trp | 3.77 |
| His | 0.87 | Tyr | 2.76 |
| Ile | 3.15 | Val | 1.87 |

More geometrically, the symmetric real and imaginary parts of the eigenvalues of $z \subset \Delta F(u)$ generate an $\alpha$-helical spiral.

$$\dot{z} = Az; \ A = \lambda t; \ \lambda = (\alpha^{-1} + i\,\alpha\omega)t \ , \tag{12}$$

with an enthalpy-entropy power law in the neighborhood of

$$\frac{\alpha^{-1}}{\alpha} = \frac{3/2}{2/3} = -2.277 \ , \tag{13a}$$

or

$$\exp^{-1.618 - 0.618} = \exp^{-2.24} \ . \tag{13b}$$

This scaling relation for the prime macromolecular hydrophobic modes can be found reflected as an aggregate dynamic of brain membrane proteins in the fundamental frequency bands of time-varying voltage oscillations of the human electroencephalogram in Hz: $\Delta = 2$ to 3; $\theta = 5$ to 7; $\alpha = 11$ to 13; and $\beta = 17$ to 19 along with their $\omega^{-\alpha}$ power law. Higher frequency, non-prime resonances are also observed, most prominently 40 Hz. These near-$\delta$ modes are superimposed on an excess-noise power spectrum, roughly $\omega^{-\alpha}$ in character; its precise enthalpy-entropy scaling exponent has not been determined. It is possible that the singularities are $\Phi n$ and the noise is $F n$, and they may scale similarly (see (17) below).

## II. Hyperbolic Dynamics of the Hydrophobic Free Energy of Proteins

$H(u)$ calculated from x-ray crystallography of protein surfaces yields $\sim 25$ to 33 cal/mol at 25 $^\circ$ C[13,14]; estimates of chain entropy loss with folding, $S(u)$, are $\sim 2$ to 3 kcal/mol/$\mathring{A}^2$ [14]; only 2 to 3 $H^+$ bonds of $\sim 600$ at $\sim 5$ kcal/mol/$H^+$ are interchangeable between solvent and self due to the relatively high specificity of their bonding sites[17]. The hyperbolic stability margin of well studied proteins like lysozyme and ribonuclease is $\sim 30$ kcal/mol [2]. Approximately 300 kcal/mol of stabilizing and destabilizing $u$ constitute a delicate balance between two large terms of opposite sign. For proteins in solution across multiple hyperbolically stable states,

$$< \Delta F(u)>^2 \, = \, < \Delta H(u)>^2 - \alpha < \Delta S(u)>^2 \sim \sqrt{u} \; + \epsilon \sim 30 \text{ kcal/mol.} \quad (14)$$

Relation (14) decomposes the phase flow of $u$, $\phi^t{}_u$, into invariant tangent spaces of the invariant **stable**, $H(u)$, and **unstable**, $S(u)$, submanifolds of $\phi^t{}_u(z)$: $z^s \otimes z^u$. $z \in \mathbf{C}$. $\phi^t{}_u$ is hyperbolic since none of its eigenvalues have norm one. $\phi^{t+s} = \phi^t \otimes \phi^s$. $\phi_0$ is the identity; $\mathbf{R}^2$, the cover on $\mathbf{C}$.

$$\phi^t{}_u : \mathbf{R}^2 \to \mathbf{R}^2 \, , \quad (15a)$$

$$\phi^t{}_u : e^{At} \, , \quad (15b)$$

$$A : \mathbf{R}^2 \to \mathbf{R}^2 \, . \quad (15c)$$

The set of primes is an open set of periodic points and the conservation of $u$, det $A = |1|$, places $\phi^t{}_u$ in the family of Anosov maps of the plane of integer-valued $n$ [18]; geometrically, the family of geodesics on a surface of negative curvature[19]:

$$A = \begin{pmatrix} n & n-1 \\ n+1 & n \end{pmatrix} \quad \text{or} \quad \begin{pmatrix} n & n-1 \\ 1 & 1 \end{pmatrix} . \quad (16)$$

We have devised another set of $\phi^t{}_u$ consonant with the above assumptions, motivated by the need to relate the behavior of these maps to the near-critical behavior of a two-parameter differential equation in $u$ to be developed in Section III. We shall call this set of one-parameter group of transformations $\mathbf{F}n$, the Farey-Anosov family, and relate the partition composed of $n\,(\mathbf{F}n)$, the number of paired quasiperiodic closed geodesics on multiple, identified copies of $\mathbf{R}^2$, to the enthalpy-compensated entropy of $F(u)$.

The Farey numbers of order $n$ are the ascending series of irreducible fractions between 0 and 1 whose denominators do not exceed $n$. If $p/q$ and $r/s$ are successive terms of the sequence, then $qr - ps = 1$. Examples of the proposed family of Farey-Anosov maps, $\phi^t{}_u = e^{At}$, are

$$A = \mathbf{F}n = 3 \tag{17a}$$

$$\begin{pmatrix} 3 & 2 \\ 1 & 1 \end{pmatrix} \begin{pmatrix} 2 & 3 \\ 1 & 2 \end{pmatrix} \begin{pmatrix} 3 & 1 \\ 2 & 1 \end{pmatrix} ,$$

$$A = \mathbf{F}n = 5 \tag{17b}$$

$$\begin{pmatrix} 5 & 4 \\ 1 & 1 \end{pmatrix} \begin{pmatrix} 4 & 3 \\ 1 & 1 \end{pmatrix} \begin{pmatrix} 3 & 5 \\ 1 & 2 \end{pmatrix} \begin{pmatrix} 5 & 2 \\ 2 & 1 \end{pmatrix} \begin{pmatrix} 2 & 5 \\ 1 & 3 \end{pmatrix} \begin{pmatrix} 5 & 3 \\ 3 & 2 \end{pmatrix} \begin{pmatrix} 3 & 4 \\ 2 & 3 \end{pmatrix} \begin{pmatrix} 4 & 5 \\ 3 & 4 \end{pmatrix} ,$$

$$A = \mathbf{F}n = 7 \tag{17c}$$

$$\begin{pmatrix} 7 & 6 \\ 1 & 1 \end{pmatrix} \begin{pmatrix} 6 & 5 \\ 1 & 1 \end{pmatrix} \begin{pmatrix} 5 & 4 \\ 1 & 1 \end{pmatrix} \begin{pmatrix} 4 & 7 \\ 1 & 2 \end{pmatrix} \begin{pmatrix} 7 & 3 \\ 2 & 1 \end{pmatrix} \begin{pmatrix} 3 & 5 \\ 1 & 2 \end{pmatrix} \begin{pmatrix} 5 & 7 \\ 2 & 3 \end{pmatrix} \begin{pmatrix} 7 & 2 \\ 6 & 1 \end{pmatrix}$$

$$\begin{pmatrix} 2 & 7 \\ 1 & 4 \end{pmatrix} \begin{pmatrix} 7 & 5 \\ 4 & 3 \end{pmatrix} \begin{pmatrix} 5 & 3 \\ 3 & 2 \end{pmatrix} \begin{pmatrix} 3 & 7 \\ 2 & 5 \end{pmatrix} \begin{pmatrix} 7 & 4 \\ 5 & 3 \end{pmatrix} \begin{pmatrix} 4 & 5 \\ 3 & 4 \end{pmatrix} \begin{pmatrix} 5 & 6 \\ 4 & 5 \end{pmatrix} \begin{pmatrix} 6 & 7 \\ 5 & 6 \end{pmatrix} ,$$

since

$$\det\,(\xi - A) = (1 - \lambda+)(1 - \lambda-) , \tag{18a}$$

$$\det\,(\xi - A) = \lambda^2 - (\mathrm{Tr}\ A)\lambda + \det A = 0 , \tag{18b}$$

$$\det A = (\lambda-, \lambda+) = 1 , \tag{18c}$$

$\dot{d}\,(\lambda-, \lambda+) > 0$ defines a divergence of $z^s \otimes z^u$ as conjugate quadratic roots of the characteristic equations of the $2 \times 2$ matrices of $\mathbf{F}n$. A symmetrical, distance-from-the-unit-circle function which grows as the maximum of $\mathrm{Tr}\ A$ will be used as an entropy-equivalent potential for relaxation of the correlation function, $C_u(\tau)$, discussed in the context of the differential equation in $u$ (see (26) below):

$$\dot{d}\left(\lambda-,\,\lambda+\right) = f(\mathbf{F}n\,) > 0 \sim C_u\left(\tau\right) < \infty \ . \tag{19}$$

The eigenvalue symmetry of $\phi^t{}_u$ predicts with increasing $\mathbf{F}n$ a greater distance from a loss of quasiperiodicity via mode coherence:

$$\dot{d}\left(\lambda-,\,\lambda+\right) = 0 \sim C_u\left(\tau\right) = \infty \ . \tag{20}$$

Since $F(u\,)$ is the source of this enthalpy-entropy compensated potential for the relaxation of $C_u\left(\tau\right)$, we begin to see more clearly how $S(u\,)$ dominates $F(u\,)$ in the hydrophobic hyperbolic dynamics of proteins.

To derive the critical $\alpha$ in the context of Anosov flows on $\mathbf{R}^2$, we define a singular condition on a $\left(z^s,\,z^u\right)$ decomposition of $\phi^t{}_u$, in which $\mathbf{R}^2$ is now a cover on $\mathbf{T}^2$, multiple copies of the unit square with points identified with those that differ by integers. Let $u = \left(u_1,\,u_2\right)$ indicate these coordinates on the space of the one-parameter group of transformations, $\phi^t{}_u$, $\mathbf{T}^2 \times \mathbf{R}^1$, now as a mapping in $u$ and $v$,

$$\phi^t{}_u = \mathbf{A}u \ , \tag{21a}$$

$$\phi_v^t = \gamma v + \cos u_1 \ , \tag{21b}$$

in which $\det \mathbf{A} = |1|$ and $\lambda_i \in \mathrm{Tr}\ \mathbf{A}$ are outside the unit circle. $v = f(u\,) : \mathbf{T}^2 \to \mathbf{T}^2$ where $f(\mathbf{A}u\,) - \gamma f(u\,) = \cos 2\pi\, u_i$. By writing

$$F(u\,) = \sum_n^\infty \gamma^n \cos\left(\lambda^{-n} t\right) \ , \tag{22}$$

we can put the Anosov flow on $\mathbf{T}^2$ (21) in the context of Hardy's treatment of the Weierstrass cosine series[20] in which an upper bound on topological conjugacy was proved to be

$$\frac{\log 1/\gamma}{\log \lambda} = 1 \ . \tag{23}$$

We note that $\exp^{\alpha^{-1}-\alpha} = \exp^{-1}$ iff $\alpha = 0.618\ldots$ In the context of Anosov maps and flows this is known generally as the Hölder continuity condition[18]. This limit can also be recovered in the symmetric scaling of the enthalpy-entropy, $\alpha^{-1}/\alpha$, as hydrophobic modes accumulate in the exponentiation of matrix $\mathbf{A}^n$ at $\mathbf{F}n = 1$ (see below).

The growth in the number of potential hydrophobic modes, the enthalpy-compensated entropy that resists the loss of geodesics[18], can be seen in the length of the period and/or the size of its terminating integer in a continued fractional expansion of the roots of $[A^n \mid (Tr\ A)_{max}]$ with increasing $Fn$ (Table 2).

| Table 2 | | |
|---|---|---|
| $Fn$ | Quadratic Surd of Root of Characteristic Equation: $\sqrt{m}$ | Periodic $(\overline{\cdots})$ Continued Fraction Expansion |
| 2 | 5 | $< 2,\overline{4}>$ |
| 3 | 12 | $< 2,\overline{2,6}>$ |
| 4 | 21 | $< 4,\overline{1,1,2,1,1,8}>$ |
| 5 | 32 | $< 5,\overline{1,1,1,10}>$ |
| 6 | 45 | $< 6,\overline{1,2,2,2,1,12}>$ |
| 7 | 59 | $< 7,\overline{1,2,7,2,1,14}>$ |
| 8 | 77 | $< 8,\overline{1,3,2,3,1,16}>$ |
| 9 | 96 | $< 9,\overline{1,3,1,18}>$ |

The general argument that $F(u) \sim S(u) \sim Fn$ leads to a prediction that the higher the average hydrophobicity of a protein per residue in kcal/mol, $< Fu >$, the more rationally independent modes it will have, and the more space it will occupy in a two-dimensional, stroboscopic x-ray crystallographic snapshot of its dynamics, $S(u) \sim dim_x$. We have calculated $< Fu >$ for three well-studied proteins with values for $dim_x$ at the extremes[21] (Table 3).

| Table 3 | | |
|---|---|---|
| | $< Fu >$ | $dim_x$ |
| Myoglobin | 1.34 | 1.67 |
| Thioredoxin | 1.38 | 1.72 |
| Trypsin | 1.19 | 1.34 |

A similar reflection of the approximations among $F(u)$, $Fn$, and $C_u(\tau)^{-1}$ in terms of enthalpy-compensated entropy may be possible in the context of species differences in the wavelength of the carboxy hemoglobin at pH 6 from time-resolved resonance spectroscopy of geminate recombination $\leqslant 10$ nsec after photo-dissociation[22]. The two examples for which amino acid sequences are available[22]

(Table 4) are consistent with the theory.

| Table 4 | | |
|---|---|---|
| $<Fu>$ | $C_u(\tau) \sim$ | $[\gamma(\text{Fe}-\text{His})\text{cm}^{-1}]$ |
| Carp | 1.38 | 216 |
| Hb-A | 1.23 | 226 |

The hypothesis from Tables 3 and 4 is less clear in the case of two of the best studied proteins (Table 5). In Section III we deal with the mechanism of that disparity.

| Table 5 | | |
|---|---|---|
| | $<Fu>$ | $\dim_z$ |
| Lysozyme | 1.15 | 1.69 |
| Ribonuclease | 1.01 | 1.33 |

III. Hydrophobic Entropy and Resistance to the Loss of Geodesic Orbits in a Hydrophobic Free Energy Hierarchy

That $<Fu> \sim Fn \sim [C_u(\tau)]^{-1}$ relates the $Fn$ partition $(z^s, z^u)$ of a hypothetical global mode of a protein to its $\dim_z$. More realistically, a protein is a hierarchy of hydrophobic modes, $z_i$, which will be represented by a geodesic pair $(z_i, z_j)$. Their rational independence, $\dot{d}(z_i, z_j) > 0$, is the system's equivalent of the Farey-Anosov condition, $\dot{d}(z^s, z^u) > 0$, for one of its modes. Whereas $\dot{d}(z^s, z^u) = 0$ implies $C_u(\tau) = \infty$, $\dot{d}(z_i, z_j) \to 0$ indicates loss of geodesic orbits through mode coherence. A polypeptide hormone's mode message $z_i$ locking onto a membrane receptor's mode $z_j$, $\dot{d}(z_i, z_j) \to 0$, generates its coherent action, $\dot{d}(z^s, z^u) \to 0$, as the mechanism of information transport in polypeptide-protein systems. Thus, hierarchical hydrophobic geodesic mode stability is involved in both internal and external macromolecular relationships, and we use a specific differential equation to examine this property.

The problem of enthalpy-compensated entropy resistance to mode loss via coherence, a reduction in $\mathbf{F}n$, can be investigated in the context of the two-parameter standard map of the circle,

$$f_{r,\lambda}(u) = u + \lambda + r/2\pi \sin(2\pi u) , \qquad (24)$$

constructed as a cross section of $T^2$ or the dissipative limit of the annulus on $\mathbf{R}^2$. Examining its critical behavior in numerical simulations requires the simultaneous manipulation of nonlinear parameter $r$ and external frequency $\lambda$. We chose instead to investigate the problem using a periodically forced nonlinear oscillator in which, except at very low $z_i$, systematic changes in $\lambda$ led to protein-relevant, internally derived arrays of $z_1, z_2, z_3 \cdots$ : a range of geodesics, $z_i \subset \mathbf{F}n$, rather than a single externally determined pair. Differentiating[23],

$$\ddot{u} = r(1 - u^2)\dot{u} + u = r\lambda \cos \lambda t \qquad (25)$$

and rescaling $(v'/r \rightarrow v)$ and noting the adjustable time step, $\Delta t = \theta/n$, we studied numerically

$$\dot{u} = v - r(u - u^3/3) , \qquad (26a)$$

$$\dot{v} = -u/r + \lambda \cos \lambda t , \qquad (26b)$$

and found examples of its sets of geodesics, $\mathbf{F}n$, on compact sets of negative curvature and instances of resonance-induced geodesic orbit loss. The dynamics emerge from two hyperbolic fixed points, saddle-sinks $p_i$ and $p_j$. Figure 1 is sketched from oscilloscopic displays of (26).

Topological conjugacy is maintained in (26) by the asymmetry in the eigenvalues of $p_i$ and $p_j$. After the orbits slow down for relatively long times in these neighborhoods, they are swept past the saddle sinks, making persistent backsided homoclinic tangencies along their insets and generating a denumerable multiplicity of geodesic orbits consistent with findings[23] concerning the $\leqslant 1$ transverse limit capacities of the saddle nodes of the "wild hyperbolic sets" of Newhouse[24].

These meromorphic functions $[f(z_i), f(z_j)] \subset \Omega \in \mathbf{C}$ are analytic in $\Omega$ except at $p_i$ and $p_j$. Since $f(z_i)$ and $f(z_j)$ have neither $z_0$ nor $z_\infty$ as poles, the $z_i$'s are bounded away from zero and infinity by the large values of $H(u)$ and $S(u)$ of opposite sign. $p_i$ and $p_j$ are the essential singularities of the Weierstrass theorem which states that analytic functions come arbitrarily close to any complex value in every neighborhood of an essential singularity[25].

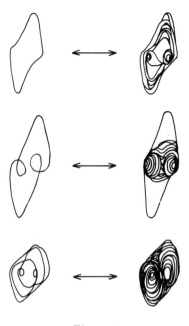

Figure 1

Loss in the number of geodesics, $\mathbf{F}n$, accompanies increases in the value of mode-coupling parameter $r$. Studies in progress[26] give a picture of this process in which $n\,(\mathbf{F}n\,)$ indicates the number of geodesics; $m\,(\mathbf{F}n\,)$, the size of the largest in $(r\,,\,\lambda)$ parameter space in which $\lambda \in [0,1]$ (Table 6).

| Table 6 | | |
|---|---|---|
| $r$ | $n\,(\mathbf{F}n\,)$ | $m\,(\mathbf{F}n\,)$ |
| . | . | . |
| . | . | . |
| 0.62 | 43 | .07 |
| 1.27 | 32 | .23 |
| 2.63 | 18 | .47 |
| . | . | . |
| . | . | . |

At $r = r_c$,

$$\phi^t{}_u = e^{At} \text{ and } \phi^t{}_u \otimes \phi^t{}_v = e^{At} , \tag{27a}$$

$$A = Fn = 1 , \tag{27b}$$

$$\begin{array}{ccc} & z_i & z_j \\ z_i & 1 & 1 \\ z_j & 1 & 0 \end{array} \tag{27c}$$

and using a symbolic dynamic subshift of finite type, the orbits accumulate like Tr A of $A^n$.

$$\begin{pmatrix} 1 & 1 \\ 1 & 0 \end{pmatrix} \begin{pmatrix} 1 & 1 \\ 1 & 2 \end{pmatrix} \begin{pmatrix} 1 & 2 \\ 2 & 3 \end{pmatrix} \begin{pmatrix} 2 & 3 \\ 3 & 5 \end{pmatrix} \begin{pmatrix} 3 & 5 \\ 5 & 8 \end{pmatrix} . \tag{28}$$

The $z_i$, $z_j$'s grow like the low order primes of the universal spectrum[27,28], $d(z_i, z_j) > 0$, and the ratio of the orbits, $z_i / z_{i+1} \sim \alpha \sim 0.618...$ The singularity at the limit of an $r$-dependent contraction in $Fn$ with det $A = |1|$, most resistant to loss of geodesics at high nonlinearity, explains the poor relationship between $< Fu >$ and $\dim_z$ in Table 5. Fourier transformation of the amino acid hydrophobic free energy sequences of these proteins (Table 1) demonstrated their dominant modes, $z_i$ (Table 7).

| Table 7 | | | | |
|---|---|---|---|---|
| | $< Fu >$ | $z_i$ | $\alpha$ | $\dim_z$ |
| Lysozyme | 1.15 | 2.2, 3.3 | 2/3 | 1.69 |
| | | 5, 7, 13.3 | | |
| Ribonuclease | 1.01 | 5.7, 10.1 | 1/2 | 1.33 |

It appears that the hydrophobic free energy geodesics of $Fn$ can be protected by an $H(u)$-compensated $S(u)$ that derives from an adequate supply of either hyperbolic hydrophobic free energy, $< Fu >$, or hydrophobic modes that are near-primes.

With respect to external communication, hydrophobic free energy mode coherence between polypeptide hormones and membrane receptor proteins has been most clearly demonstrated in a pituitary cell perifusion system for six peptides that release ACTH with equal potency[29] yet have less than 50% homology in their amino acid sequences; all have a dominant $z_i$ of 2.1 residues[30]. ACTH itself has a $z_i$ of 2.2, and so does its membrane receptor protein[31]. This generates a polypeptide-membrane protein information transport cascade:

$$\left( \left[ \dot{d}\,(z_i,\,z_j) \to 0 \right] \to \left[ \dot{d}\,(z^{\,s},\,z^{\,u}) \to 0 \right] \right) \cdots . \tag{29}$$

It may be relevant that actions of components of proopiomelanocortin, the brain and pituitary peptide precursor, (activation of stress hormones, stimulation, sedation, analgesia) are associated with a complete supply of near-prime hydrophobic modes of characteristic residue wavelengths: 2.2, 3.6, 5, 6.67, 10.5, 13.3, and 20[9].

The supply of primes as the entropy that protects against the loss of quasi-periodic geodesics, its relationship to the unique Hölder continuity exponent (23) at the symmetric enthalpy-entropy fixed point $\alpha^{-1}/\alpha$, and the map of (24) at $r = r_c$ are brought together in a recent unpublished numerical study[32]. All three generic, two-parameter maps of the circle yielded the same "mysterious" power law at $r = r_c$ when relating $\log m\,\mathbf{F}n$ to $\log n\,\mathbf{F}n$ as normalized by the Euler prime function, $\Phi n : \exp^{-2.3}$ (13).

We are studying (25) and (26) to see if this universal power law at $r = r_c$ holds in the more natural context of a two-parameter nonlinear differential equation in $u$. It is likely that it does because the original study of the equation[23] demonstrated that $\dot{d}\,(z_i,\,z_j) \to 0$ at mod $|z_i - z_j| = 2/3$ mod $z_j$.


IV. How Proteins Work

With the high spatial resolution of mega-ohm voltage clamp techniques, time-dependent ion conductances in resting allosteric nicotinic cholinergic membrane receptor proteins manifest multiple conductance states that differ in amplitude and duration, and each contains excess noise[33]. Cholinergic neurotransmitter-like ligands convert this pattern of multiple concentric time and space scales to a pair of more widely spaced length and time scales: bursting in microseconds and interburst intervals in milliseconds[34]. The germ of the fast activity may have been present in the excess noise of the resting condition. One view of this signal-induced transition is as a tangent bifurcation of the universal spectrum, $\mathbf{F}n = 1$, where the multiple fractal barrier-height or deep-trap diffusivity[35] going like $t^{-\alpha}$ ($0 < \alpha < 1$) crosses through a $t \ln t$ region of the distribution of prime orbits to acceleration in ion diffusion like $t^2$ [36], or even to a turbulent mixing transport regime, $t^3$ [37]. Like bursting itself, living things are precipitates of accelerated physical time, and the common mechanism may be tangent bifurcation to intermittent contraction and turbulent diffusion in the hydrophobic free energy of proteins. Increasing intramolecular pressure during the shrinking of length scale with bursting could overcome short range repulsive resistances to catalysis and binding; turbulent mixing may transcend the rate-limiting linear-in-$t$ or $t^{-\alpha}$ diffusivities ($0 < \alpha < 1$).

We suggest that proteins participate actively in binding, catalysis, and transport via low hydrophobic free energy phase transitions among their multiple states of enthalpy-compensated entropy to more efficient bursting dynamics. In contrast, some theories apply the Kramers equation[38] to passive macromolecules and offer a picture of Brownian motion along a one-dimensional reaction coordinate composed of a sequence of potential energy barriers. They conjecture that random solvent collisions at $37\,^{\circ}$C generate energy subsequently dissipated by viscous dampening, a fluctuation-dissipation theory of the source of energy by which chemical processes surmount barrier heights along the reaction coordinate[39]. Studies that have been used to confirm this linear diffusion theory cite evidence that increasing solvent viscosity reduced catalytic and geminate rebinding reaction rates[40,41]. An alternative interpretation is that since the chemicals used to augment solvent viscosity were mono, polyhydroxy organic compounds or comparable phosphate buffers, they reduced the surface tension-like hyperbolic hydrophobic free energy of the solvent-macromolecule interface.

In the context of (24) and (26), the transition to intermittent turbulent motion via homoclinic tangency in hyperbolic systems dominated by two diffeomorphic saddle node fixed points occurs naturally at the boundary of the set of rationally independent rotations, as in (27)[42]. Accelerated diffusion of points around two destabilized fixed points with eigenvalues in the close vicinity of the unit circle, $(\lambda-, \lambda+) \in (z^s, z^u)$, was first reported in a transformation of the Lorenz system into a generic nonlinear oscillator as in (26)[43, Fig. 1]. As a lift to the unit interval, the paracritical region for (24), $r \sim r_c$, has been explored using a cross section and then an $x_n/x_{n+1}$ mapping which bounces back and forth down the corridor formed by a parabolic curve and its 45° bisectrix, which the function first intersects, is tangent to, and then loses contact with as $r \to r_c$[44]. The scenario of hydrophobic free energy transitions in proteins moves in the opposite direction.

A broader theoretical issue involves the distribution of singularities in non-linear biological systems and whether renormalization group transformations are necessary to elucidate their mathematical physics. Is it possible that renormalization in protein hydrophobic free energy dynamics simply means shifting the logarithmic bases of the process to an appropriate range of the nonlinear coupling-dependent supply of available $Fn$? We know from one-dimensional mappings of the unit interval[45] that the limit on stability of $\phi^t_u$ is that $\dot{\phi}^t_u < |1|$ as in (23); $\dot{\phi}^t_u = -1$ indicates a nascent period-doubling bifurcation, and $\dot{\phi}^t_u = +1$ indicates incipient intermittent bursting. The allosteric protein appears to rest in a state near these two possibilities at all scales due to its uniform hyperbolicity condition. Farey-Anosov maps and flows, $Fn$, det $A = |1|$ constitute a discontinuous continuum of a denumerable infinity of states in the form of a distribution, $Fn$, resembling that of the primes[46]. With the zeta function, the density distribution of these geodesics on a surface of negative curvature, $Fn$, has been proved to be analogous to that of the prime numbers[47,48]. The underlying supply of singularities in both analytic and geometric number systems may account for the recurrent oscillatory poles in renormalization group transformations, periods depending on logarithmic length scale[49,50]. This approach substitutes the concept of a pre-existing field of numbers that can be gathered into hyperbolic fixed points by nonlinear coupling for the renormalization idea that fixed points generate the field.

X-ray crystallographic studies of proteins demonstrate that fewer than five percent of their 125 to 150 amino acids are involved in site-specific binding or catalysis. The functional role of the other amino acids is a continuing area of investigation. Another, more **fundamental**, unknown concerns the source and mechanisms of storage and signal-specific release of energy used in catalytic, binding, and transport functions, a thermodynamic basis for living things. The major thrust of our work concerns how it is that the first mystery serves as the answer to the second.

Among the implications of a hydrophobic-hyperbolic free energy theory of protein structure and function are the following.

1. Since RNA codons for the most hydrophobic amino acids have uracil in position 1 or 2[51] and hydrophobically equivalent amino acid substitutions leave tertiary structure and dynamics intact[52], there may be a more general scheme for macromolecular information than specific DNA-RNA-amino acid coded sequences. For example, whereas a large number of hydrophobically equivalent amino acid mutations in hemoglobin leave its oxygen binding kinetics intact, substitution of valine for glutamate (0.67 for 1.87 kcal/mol of hydrophobic free energy) is associated with the defects of sickle cell anemia[53].

2. Pathologically persistent bursting behavior is characteristic of a number of medical disorders; cardiac arrhythmias, pituitary tumors, and some epilepsies serve as examples. A pharmacology of these and similar dynamical disorders may involve the development of agents that increase or decrease the hyperbolic hydrophobic free energy of proteins, a pharmacology of global dynamics rather than function-specific sites.

## References

1. Stillinger, F.H. *Science* 209: 451, 1980.
2. Privalov, P.L. and Khechinashvili, N.N. *J. Mol. Biol.* 86: 665, 1974.
3. Tanford, C. *Proc. Natl. Acad. Sci. USA* 76: 4175, 1979.
4. Teipel, J. and Koshland, D.E. *Biochemistry* 8: 4656, 1969.
5. Levitzki, A. and Koshland, D.E. *Proc. Natl. Acad. Sci. USA* 62: 1121, 1969.
6. Mandell, A.J. and Russo, P.V. *J. Neurosci.* 1: 380, 1981.
7. Russo, P.V. and Mandell, A.J. *Brain Res.* 299: 313, 1984.
8. Aubry, S. *Physica D* 7: 340, 1983.
9. Mandell, A.J. *Annu. Rev. Pharmacol. Toxicol.* 24: 237, 1984.
10. Antonini, E. and Brunori, M. *Hemoglobin and Myoglobin in their Reaction with Ligands.* North-Holland, Amsterdam, 1969.
11. Ruelle, D. *Thermodynamic Formalism.* Encyclopedia of Mathematics, Vol. 5. Addison-Wesley, Reading, MA, 1978.
12. Gnedenko, B.V. and Kolmogorov, A.N. *Limit Distributions for Sums of Independent Random Variables.* Addison-Wesley, Reading, MA, 1968.
13. Reynolds, J.A., Gilbert D.B. and Tanford, C. *Proc. Natl. Acad. Sci. USA* 71: 2925, 1974.
14. Chothia, C. *Nature* 248: 338, 1974.
15. Janin, J. *Bull. Inst. Pasteur* 77: 337, 1979.
16. Nozaki, Y. and Tanford, C. *J. Biol. Chem.* 246: 2211, 1971.

277

17. Kauzmann, W. *Adv. Prot. Chem.* 14: 1, 1959.
18. Cornfield, I.P., Fomin, S.V. and Sinai, Y.G. *Ergodic Theory.* Springer-Verlag, Berlin, 1980.
19. Anosov, D.V. *Sov. Math.* 3: 1068, 1962.
20. Hardy, G.H. *Trans. Am. Math. Soc.* 17: 301, 1916.
21. Wagner, G.C., Colvin, J.T., Allen, J.P. and Stapleton, H.J. *J. Am. Chem. Soc.* in press, 1985.
22. Friedman, J.M. *Science* 228: 1273, 1985.
23. Cartwright, M.L. and Littlewood, J.E. *J. Lond. Math. Soc.* 20: 180, 1945.
24. Palis, J. and Takens, F. Hyperbolicity and the creation of homoclinic orbits. I.H.E.S. preprint, 1985.
25. Ahlfors, L.V. *Complex Analysis.* McGraw-Hill, New York, 1979, pp. 129.
26. Mandell, A.J., Russo, P.V. and Woyshville, M. ms. in preparation, 1985.
27. Shenker, S.J. *Physica D* 5: 405, 1982.
28. Rand, D., Ostlund, S., Sethna, J. and Siggia, E.D. *Phys. Rev. Lett.* 49: 132, 1982.
29. Rivier, J., Rivier, C. and Vale, W. *Science* 224: 889, 1984.
30. Mandell, A.J. in *Synergetics of the Brain* (eds. Basar, E., Flohr, H., Haken, H. and Mandell, A.J.). Springer-Verlag, Berlin, 1983.
31. Bost, K.L., Smith, E.M., and Blalock, J.E. *Proc. Natl. Acad. Sci. USA* 82: 1372, 1985.
32. Lanford, O.E. A numerical study of the likelihood of phase locking. I.H.E.S. preprint, 1985.
33. Sachs, F. in *Single-Channel Recording* (eds. Sakmann, B. and Neher, E.). Plenum, N.Y., 1983, pp. 365.
34. Colquhoun, D. and Sakmann, B. *Nature* 294: 464, 1981.
35. Montroll, E. and Shlesinger, M. in *Nonequilibrium Phenomena II.* North-Holland, Amsterdam, 1984, pp. 1-121.
36. Geisel, T., Nierwetberg, J. and Zacheril, A. *Phys. Rev. Lett.* 54: 616, 1985.
37. Shlesinger, M.F. and Klafter, J. Lévy walks versus Lévy flights. *Proceedings, Cargese Summer School on Fractal Growth,* in press, 1985.
38. Kramers, H.A. *Physica* 7: 284, 1940.
39. Gavish, B. *Biophys. Struct. Mech.* 4: 37, 1978.
40. Gavish, B. and Werber, M.M. *Biochemistry* 18: 1269, 1979.
41. Beece, D., Eisenstein, L., Frauenfelder, H., Good, D., Marden, M.C., Reinisch, L., Reynolds, A.H., Sorenson, L.B. and Yue, K.T. *Biochemistry* 19: 5147, 1980.
42. Newhouse, S., Palis, J. and Takens, F. *I.H.E.S.* 57: 1983.
43. Morioka, N. and Shimizu, T. *Phys. Lett. A* 66: 447, 1978.
44. Manneville, P. and Pomeau, Y. *Physica D* 1: 219, 1980.
45. Feigenbaum, M.J. *Physica D* 7: 16, 1983.
46. Hardy, G.H. and Wright, E.M. *An Introduction to the Theory of Numbers.* Oxford Univ. Press, Oxford, 1979.
47. Manning, A. *Bull. Lond. Math. Soc.* 3: 215, 1971.

48. Parry, W. and Pollicott, M. *Ann. Math.* 118: 573, 1983.
49. Nauenberg, M. *J. Phys. A* 8: 925, 1975.
50. Shlesinger, M.F. and Hughes, B.D. *Physica A* 109: 597, 1981.
51. Dickerson, R.E. and Geis, I. *The Structure and Actions of Proteins.* Benjamin, Reading, MA, 1969.
52. Lesk, A.M. and Chothia, C. *J. Mol. Biol.* 136: 225, 1980.
53. Watson, J.D. *Molecular Biology of the Gene.* Benjamin, Reading, MA, 1977.

# Remarks about Nonlinear Dynamics

## John Guckenheimer

Nonlinear dynamics as a field of scientific research has its theoretical and applied aspects. On the one hand, much is known about the typical or generic behavior of dynamical systems and their bifurcations, and more is being discovered by theoretical and computational investigations. On the other hand, a methodology has developed which can be employed as a toolkit for the study of a wide range of physical phenomena. The future of the subject involves both the theoretical and applied aspects and is dependent on their interaction for its continued vitality. These remarks outline briefly my perception of what are currently outstanding issues and discuss their significance.

An important aspect of attractors is their information dimension. The information dimension is related to the function which describes the proportion of time that a typical trajectory spends within a given distance of a typical point within the attractor. If this function has a power law dependence on the distance as it tends to zero, the exponent gives the information dimension. It provides a significant measure of the size of the attractor in terms of the recurrence time for trajectories. In many experimental studies it is difficult to prepare precise initial conditions on an attractor. Thus the only way to follow the evolution of nearby trajectories may be to wait for a previously observed state in the attractor to recur (approximately) in the subsequent evolution of a single trajectory. The recurrence time increases exponentially with the information dimension, so there are practical limitations on the study of the details of attractors of large or even moderate information dimension. Thus measurements of the information dimension provide a diagnostic tool for determining whether it will be feasible to explore the details of the dynamics within an attractor.

The issue of how one can study attractors of moderate information dimension is certainly one of the outstanding issues of nonlinear dynamics. From a naive point of view, it seems that the regimes governed by low dimensional attractors are likely to be rather small in the parameter spaces of individual problems. In dealing with attractors of moderate dimension, one will not be able to produce high resolution pictures that resolve the structure of the attractor on small scales. It seems that the more probabilistic ideas of ergodic theory provide the only tools for studying such attractors. There is considerable scope here for both theoretical and experimental work. The existence

of the information dimension and its relation to the ideas of
entropy and characteristic exponents have been substantially
clarified in the past couple of years. These aspects of the
subject are still rapidly evolving. It may be a few more years
before a fully mature view of these ideas is at hand. The
statistical robustness and stability of techniques for
estimating information dimension of attractors are a topic
of considerable practical importance. For applications to
fluid mechanics, one would like to explore the relationship
between observed spatial structures within a fluid and the
dimension of phase space attractors. Are high dimensional
attractors an amalgamation of weakly coupled coherent
spatial structures which persist for reasonably long
periods of time?

A second major issue in nonlinear dynamics involves
studying the bifurcations that describe how attractors
change qualitatively as parameters are varied. This is
an area which has been greatly enriched by the use of
renormalization methods and ideas about scaling. Experimental
observations based upon bifurcation theory have been
largely responsible for the public attention that nonlinear
dynamics has received in the past few years. The models
provided by bifurcation theory have made very successful
predictions for diverse physical and chemical phenomena.
The outstanding problems here involve systems with
several parameters and the influence of symmetry. Many
dynamical models are formulated in terms of systems of
ordinary differential equations with several variables
and several parameters. The problem of even solving the
(nondifferential) equations to find the equilibria and
their stability as a function of the parameters is a
difficult computational task. The project of building
a software package for investigating the dynamical
properties of systems of moderate size could be of great
help in many situations. The elucidation of the unfoldings
of bifurcation problems with an imposed symmetry and of
higher codimension bifurcations is an important task in
its own right and also a good test for the capabilities
of numerical software.

A third issue for nonlinear dynamics is one of finding
useful applications. The theory does provide us with tools
for understanding a range of phenomena which were previously
inaccessible to analysis. This knowledge should be of help
in design and engineering as well as in understanding basic
mechanisms involved in physical instabilities. The subject
is still too young to know where the important applications
will lie, but it is a reasonable presumption that they will
come.

# INFINITE DIMENSIONAL DYNAMICS

Jack K. Hale

Division of Applied Mathematics

Brown University

Providence, RI 02912 USA

1.    Introduction.

Many applications involve dynamical systems in infinite dimensional spaces; for example, those generated by partial differential equations and delay differential or functional differential equations. Due to the complexities involved in performing analysis in infinite dimensions, one often imposes simplifying assumptions on the model to obtain a problem in n dimensional euclidean space $R^n$ or, more generally, in a finite dimensional manifold $M^n$. These assumptions can be either physical hypotheses or some finite dimensional approximation scheme. Using the extensively developed geometric theory of dynamical systems on a finite dimensional manifold, one attempts to understand the behavior of the original system as the relevant physical parameters are subjected to variations.

Finite dimensional dynamics, especially mappings in dimension $\leq 2$, have had a tremendous impact on the present thinking in physical systems. This is especially true in the field of chaotic dynamics as can be seen by perusing the literature. It is clear that this has been a stimulus to the theoreticians as well as the experimentalists. Whether the dynamics of low dimensional spaces can describe complicated physical phenomena is still an open problem. The validity of the process of reduction to some finite dimensional space can only be justified if we have some reasonable geometric theory for infinite dimensional dynamical systems themselves. That is, we must have a theory for infinite dimensional systems which uses and appropriately modifies the concepts and ideas from finite dimensions in a significant way and, at least on a theoretical level, gives a prescription for determining the nature of the dynamics. Also, a dynamical system being infinite dimensional does not necessarily mean that it can exhibit every type of dynamics that occurs in ordinary differential equations in $R^n$. The infinite dimensional system may have special properties which must be isolated a priori and will determine the types of approximating systems that are feasible.

Although efforts are being made to extend the finite dimensional ideas to infinite dimensions, the global theory is still in its infancy. One reason the development has been so slow certainly is due to the complexities that are introduced by the infinite dimensionality. It is difficult to isolate important examples which can be analyzed in sufficient detail to bring out some of the new phenomena that occur and, at the same time, show the advantages to be gained from considering the system as an abstract dynamical system. These advantages can be either in the discovery of new types of behavior or, perhaps more importantly, the approach can lead to the formulation and analysis of much more meaningful problems.

For the successful development and application of dynamical systems in infinite dimensions, we need intensive interaction between two special groups of researchers. The first group consists of mathematicians who are well trained in dynamical systems and know both the analytic and the geometric theory of differential equations in finite dimensions. They should also know well the classical and modern theory of partial differential and functional differential equations and have a strong background in applications - especially physics and engineering. The other group of researchers should be primarily concerned with applications, but should be well trained in ordinary and partial differential equations. It does not take much reflection to see that there are very few people with these qualifications. More resources need to be allocated for training young people to carry out this program.

In the past thirty years, there have been several instances where the type of interaction above has led to considerable success. A few illustrations will perhaps convey the spirit of the remarks above.

## 2. Functional differential equations.

The importance of delay differential or functional differential equations (FDE's) in viscoelasticity, mathematical biology and control theory has been emphasized for a long time (Volterra[36),37),38)], Minorsky[28)] and references therein). The early method of attack on such a problem often was to assume the system either was linear so that Laplace transform techniques could be applied or to make severe hypotheses that would permit the direct application of techniques from ordinary differential equations. Mishkis[29)], Bellman and Cooke[2)] are excellent references for the beginning of this theory - especially linear systems. Although Volterra[37] had employed techniques similar to Liapunov functionals, it was Krasovskii[19] who emphasized the importance of considering FDE's as dynamical systems in infinite dimensional spaces. This book, presented to me by Professor Lefschetz, was a prime mover for my own research in this area. At this time, I think we are justified in saying that the subject is now mature and the foundations of the geometric theory are complete (see Hale[10], Hale, Magalhães, and Oliva[13]). Most of the ideas from finite dimensions have been appropriately modified and extended so as to apply to FDE's.

At the present time, the subject of FDE's is part of the vocabulary of applied mathematicians and engineers. It is interesting to note that the engineering literature contained very few papers on delay equations in the 1950's, whereas it is a standard topic today. Although the space program certainly has played an important role in this change, one must also give credit to the availability of the fundamental theory in a form accessible to engineers.

As remarked earlier the subject of FDE's has matured enough to enable one to begin to study in detail the dynamics of particular systems. By doing so, some very surprising results have been obtained. To be more specific, we must become somewhat technical to make sure that the setting for the discussion is clear.

If $T(t)$, $t \geqslant 0$, is a semigroup on a Banach space $X$, the attractor $A$ of $T(t)$ consists of the nonempty set of orbits of $T(t)$ which can be defined for all $t \in R$ and are bounded. The attractor $A$ contains in particular the $\omega$-limit set of all bounded orbits and, therefore, contains all information necessary for determining the structure of the flow defined by $T(t)$. The semigroup $T(t)$ is point dissipative if there is a bounded set $B$ in $X$ such that $T(t)x \in B$ for $t \geqslant t_0(x,B)$. An important result on dissipative systems is the following: (Billotti and LaSalle [1971]) If $T(t)$ is point dissipative and there is a $t_1 > 0$ such that $T(t)$ is compact for each $t \geqslant t_1 > 0$, then the attractor $A$ is the maximal compact connected invariant set of $T(t)$, $t \geqslant 0$ and the $\omega$-limit set of any bounded set belongs to $A$. This theorem shows that the attractor $A$ exists, is compact and is very stable. Therefore, it is reasonable to restrict the discussion to the attractor $A$ of the flow defined by $T(t)$, in order to determine the significant properties of the flow. FDE's satisfy the smoothness conditions imposed in the theorem above.

A <u>Morse decomposition</u> for the compact attractor A is a finite ordered collection of compact disjoint sets $S_i \subset A$, $i = 1,2,...,N$, called <u>Morse sets</u> such that for any $x \in A$, there exist integers $i \leqslant j$ such that

$$\text{dist}(T(t)x,S_j) \to 0 \quad \text{as } t \to -\infty ,$$

$$\text{dist}(T(t)x,S_i) \to 0 \quad \text{as } t \to +\infty , \qquad (1)$$

and, if $i = j$, then $T(t)x \in S_i$ for $t \in R$.

In some sense, a Morse decomposition of A is a generalization of a gradient system. For a gradient system, the sets $S_i$ are singletons corresponding to the equilibria. In the more general case, the sets $S_i$ could be, for example, periodic orbits or even "strange attractors" which support very complicated dynamical behavior.

Mallet-Paret[25] has shown that certain classes of differential difference equations possess Morse decompositions. More precisely, a scalar equation which satisfies this property is

$$x(t) = f(x(t),x(t-1)), \qquad (2)$$

with a negative feedback condition

$$yf(0,y) < 0, \quad y \neq 0 ,$$

$$\partial f(0,0)/\partial y < 0. \qquad (3)$$

The sets $S_i$ are characterized by means of an integer-valued Lyapunov function V on $A\backslash\{0\}$ which measures the number of oscillations that a solution undergoes in a unit interval. The set $S_i$ roughly corresponds to orbits where $V = i$. The manner in which the sets $S_i$ are connected by orbits is still an open problem.

This is a very surprising result and depends very strongly upon the fact that the delay equation is a scalar and f has negative feedback. On the other hand, it also illustrates clearly how important qualitative information can be obtained by appropriately modifying ideas from classical dynamical systems.

3.    Parabolic systems.

In any study of partial differential equations, much effort is needed to obtain a basic understanding of the existence, uniqueness, continuous dependence and regularity theorems. Once this topic is reasonably well understood, one can turn to the discussion of the qualitative properties defined by the flow. In the last twenty-five years, significant progress has been made in the latter topic for parabolic equations. Some of the major contributions can be found in Sobolevski[32], Friedman[6),7)], Fujita and Kato[8], Henry[17]. The presentation of Henry [1981] (partially completed in 1972 and circulated as notes) on the geometric theory has brought the subject to the same level as the theory of FDE's and in a much shorter time span. Perhaps having the theory of FDE's as a guide assisted in achieving this goal.

The basic problem in the qualitative theory is the same as for FDE's - study the flow on the attractor and see how this flow depends on parameters.

Let us mention one striking result on a scalar parabolic equation

$$u_t = u_{xx} + f(x,u,u_x), \qquad 0 < x < 1 , \qquad (4)$$

plus homogeneous boundary conditions at  $x = 0$  and  $x = 1$ . If  $\phi$  is a hyperbolic equilibrium point, and  $W^s(\phi)$,  $W^u(\phi)$  are its stable and unstable manifolds, we then have the following theorem:

Theorem (Henry[18]).    $W^u(\phi) \pitchfork W^s(\psi)$  for all hyperbolic equilibria  $\phi, \psi$, where "$\pitchfork$" means transversal.

For ordinary differential equations, a theorem of this generality is only valid in one dimension. Yet here is an infinite dimensional parabolic equation with this property!

The proof of the theorem uses some results not only from PDE's (the maximum principle, Sturm-Liouville theory) but also employs important ideas from dynamics (properties of stable and unstable manifolds, exponential dichotomies).

One important lemma in the proof of the theorem above which perhaps is more important than the theorem itself is

Lemma (Henry[18]).    If  $\phi, \psi$  are hyperbolic equilibria and there is an orbit

<u>connecting</u> ɸ <u>to</u> ψ, <u>then</u> dim $W^u(ɸ)$ > dim $W^u(ψ)$.

The system above is gradient-like and every bounded solution approaches an equilibrium solution as t → ∞. If the function f in the parabolic equation depends only on x, u, and $\lim_{|u|\to\infty} f(x,u)/u$ < 0, then all solutions are bounded and the flow has a compact attractor with a Morse decomposition (see Henry[18]). The $i^{th}$ set $S_i$ in the Morse decomposition can be chosen to be the set of equilibria ɸ with dim $W^u(ɸ)$ = i. The results above can also be of assistance in determining the manner in which equilibrium points are connected on the attractor (see Henry[18]).

This example is another good illustration of the role that dynamical systems can play in the formulation of reasonable and important questions for infinite dimensional systems. Let us give another illustration.

For the Navier-Stokes equation, in a bounded domain Ω with Dirichlet boundary conditions, Ladyzenskaya[20] has shown that every solution (considered as element of $H_0^1(Ω)$) has its limit supremum as t → ∞ bounded by a constant which depends only on the norm of the density of force per unit area. This implies that the semigroup defined by the Navier-Stokes equation is point dissipative. Since this semigroup is also compact, this implies (by the theorem mentioned in Section 2) that there is a compact attractor A which has the property that the ω-limit set of each bounded set belongs to A; in particular, dist(T(t)B,A) → 0 as t → ∞ in $H_0^1(Ω)$. Ladyszenskaya[20],[21], also exhibited a set $M_R ⊂ B_R$, the ball of large radius R in $H_0^1(Ω)$ which is compact and consists of the solutions which are globally defined and remain in $B_R$. She then asks: does $M_R$ coincide with the ω-limit points of the Navier-Stokes equation? This is clearly not true because $M_R$ = A, the attractor above and therefore must be connected. It contains in particular orbit connections between minimal sets. The stability properties of A were not pointed out by Ladyzenskaya. Since A is also the maximal compact invariant set, this answers in the affirmative a problem posed by Temam[35] on the boundedness of A.

It has been shown by Constantin, Foias and Temam[4] that, for the Navier-Stokes equation in a bounded domain Ω in $R^3$, the concept of point dissipativeness is equivalent to the global existence of solutions. Thus, global existence is equivalent to the existence of a compact attractor.

In recent years, Prodi, Constantin, Foias, Temam, Manley, Treve, Mascke,

Saromito (for references, see [Constantin, Foias, and Temam[4]]) have been interested in the problem of finite determinancy of the flow on the attractor. In a vague sense, this means that the asymptotic properties of the solutions are determined by the asymptotic properties of the first m Fourier coefficients of the Fourier series (m large). For some problems, finite determinancy has been verified and explicit estimates on m in terms of the Reynolds number have been given.

A positive solution of the finite determinancy problem defined above does not show that a finite number of Galerkin approximations determine the properties of the flow since exact solutions are used in finite determinancy.

Using more of the dynamical properties of the flow, one can show that a finite number of Galerkin approximations do determine the attractor of the flow. In fact, if A is the attractor for the Navier-Stokes equation, then there is an $m_0$ such that, for any $m \geqslant m_0$, the Galerkin approximate equations of order m have an attractor $A_m$ that converges to A as m → ∞. This result is valid for Galerkin approximations which use Fourier series or splines (see Hale, Lin, and Raugel[12]). If one assumes, in addition, that the flow on the attractor has a gradient structure, then it can be shown that the flow defined by the Galerkin approximations has the same qualitative dynamics as the flow for the PDE. This certainly suggests that the Galerkin approximations for m large carry all of the quantitative information about the flow. Related results are contained in Foias and Guillope[5] and Mallet-Paret and Sell[25]. There are other illustrations of how restricting the discussion to the attractor permits one to ask more sophisticated and important questions. It is possible to study the effects of variations in parameters of a more complicated type; for example, variations in diffusion coefficients and boundary conditions (see Hale and Rocha[14]).

4. Hyperbolic systems.

At this time, the theory of dynamical systems, in the spirit referred to above, has not been developed very extensively for hyperbolic partial differential equations. Only a few special cases have been considered.

One of the earliest cases occurred in the theory of lossless transmission lines with nonlinear boundary conditions (the telegraph equation with nonlinear boundary conditions). In one space variable, the problem can be transformed into an equation with delayed arguments in the highest derivative. Equations of this type are called neutral functional differential equations (NFDE's). These equations also arise in other oscillatory systems with some interconnections between them; for example, vibrating masses attached to an elastic bar, the collision problem in electrodynamics, etc. For a large class of NFDE's the geometric theory is quite sophisticated and is almost at the level of FDE's and Parabolic PDE's (see Hale[10]).

For a NFDE in this class, there will be a compact attractor if the corresponding semigroup $T(t)$ is point dissipative and orbits of bounded sets are bounded (see Massatt[27]). This semigroup is not compact for any $t$, but is the sum of a strict linear contraction and a nonlinear completely continuous operator. This latter property plays an important role in the existence of the attractor and is shared by other hyperbolic systems.

Recently, Babin and Vishik[1] have extended the theory of gradient systems of Henry[17] so that it applies to hyperbolic PDE's with a linear damping term. They prove the existence of an attractor $A$ and some properties of stability for the attractor. Using the results of Massatt[27] and the fact that the semigroup defined by the equation has the property mentioned above for NFDE's, it follows (see Hale[11]) that the $\omega$-limit set of any bounded set belongs to the attractor $A$, a much stronger conclusion than in Babin and Vishik[1]. Haraux[16] has obtained similar results.

Since the solution operator $T(t)$ satisfies the properties stated above, the attractor $A$ must have finite Hausdorff dimension (Mallet-Paret[22]) and even finite capacity (Mañé[26]). For another proof, see J.-M. Ghidaglia and R. Temam[9].

If we consider $T(t)\phi$, $t \in R$, $\phi \in A$, then the results in Hale and Scheurle[15] imply that $T(t)\phi$ is as smooth in $t$ as the vector field, even up to analyticity.

A careful examination of the recent papers of Strauss and Shatah[31] on stability and instability of bound states for hyperbolic PDE's on $R^n$ indicates how ideas from dynamical systems can be of assistance in the formulation of results as well as suggesting methods of attack on solutions of problems. The recent Ph.d. thesis of Sternberg[33] deals with the same topic on bounded open sets in $R^n$.

## 5.  Chaotic dynamics.

Since the time of Poincaré, the existence of homoclinic orbits in differential equations has been known to imply the existence of complicated oscillatory phenomena. By the early 1970's, much of the basic theory of the behavior of solutions near transverse homoclinic orbits had been developed principally by Birkhoff, Smale, Sil'nikov, Gavrilov, Newhouse, and Palis.

Potential applications of the abstract theory had been suggested, but probably Ruelle and Takens[30] had the most impact on making people in the applications aware of the theory when they attempted to relate it to turbulence. Over the past few years, chaotic motion or strange attractors have been observed in many branches of mechanics, engineering and biology.

Perhaps the most important problem in this field of investigation is to understand the process through which chaos is created in a particular physical problem or differential equation. This is not only important to our basic understanding of the phenomenon, but perhaps will permit the design of appropriate controls to prevent chaos. Nonchaotic motion is certainly preferable to chaotic motion.

In an attempt to understand the mechanism for the onset of chaos, it was very natural to study the simplest type of mappings; for example, the logistic map, Henon map, the time one map for the Lorenz equation, etc. It is the hope that ideas obtained from this type of investigation could then be adapted to applications. At this time, however it seems as if the people in the applications often must resort to comparing the experimental data (pictorially) with properties that have been obtained theoretically or numerically for special mappings. We must increase our efforts to understand the mechanism for the onset of chaos in specific real world problems. One possibility on both the theoretical and applied side would be the following. Given a theory for the onset of chaos, devise a numerical procedure for testing the validity of this theory on a particular physical problem. Without such a procedure, we can never relate theoretical results to applications. We should not be content to say that a dynamical phenomenon in a very high dimension space is the same as that of a two dimensional mapping because the two dimensional pictures of the dynamics are similar. This is a very hard problem but definitely needs to be pursued. Such numerical procedures are not available even for finite dimensional problems. Infinite dimensional problems must be approximated by finite dimensional problems, but the types of "good" approximations may come from a better understanding of the infinite dimensional dynamics.

This is not to say that there are not some very interesting ideas around, but these ideas perhaps are not being exploited as much as they should be. Takens[34] (see also other references there) has proposed an interesting method for trying to determine the dimension of the attractor of dynamical systems. Usually one does not observe the instantaneous state of a dynamical system, but only a function of the state. If we observe this function of the state a sufficiently large number of times, then one would expect to recover properties of the state. This observation has been exploited by Takens[34]. It is also interesting to observe that a similar trick was used by Mallet-Paret[23] to show that, generically, periodic orbits are dense for FDE's. The observable quantity in FDE's is the velocity at time t which is a function of the state over an interval [t-r,t].

6. Future plans.

It is my hope that the illustrations above indicate that it is very worthwhile to attempt to develop a geometric theory of infinite dimensional systems. One should try in every way to see how the important ideas in finite dimensions can be adapted to the infinite dimensional setting. This process not only will clarify the problem, but will lead to the discovery of completely new and unexpected results for particular types of equations. It is my feeling that we need to devote a considerable portion of our resources to the achievement of this goal.

References

1) A. V. Babin and M. I. Vishik [1983], Regular attractors of semigroups of evolutionary equations. J. Math. Pures et Appl. 62 (1983), 441-491.
2) R. Bellman and K. Cooke [1963], Differential Difference Equations. Academic Press.
3) J. E. Billotti and J. P. LaSalle [1971], Periodic dissipative processes. Bull. Am. Math. Soc. 6(1971), 1082-1089.
4) P. Constantin, C. Foias and R. Temam [1984], Attractors representing turbulent flows. Preprint, Univ. Paris-Sud, Dep. Math. 84T35.
5) C. Foias and C. Guillopé [1985], On the behavior of the solutions of Navier-Stokes equations lying on invariant manifolds. J. Differential Equations, submitted.
6) A. Friedman [1964], Partial Differential Equations of Parabolic Type. Prentice Hall.
7) A. Friedman [1969], Partial Differential Equations. Holt, Rinehart and Winston, New York.
8) H. Fujita and T. Kato [1964], On the Navier-Stokes initial value problem. Arch. Rat. Mech. Anal. 16(1964), 269-315.

9) J.-M. Ghidaglia and R. Temam [1985], Propriétés des attracteurs associés a des equations hyperboliques non lineares amorties. C. R. Acad. Sci. Paris 300, 185-188.

10) J. K. Hale [1977], Theory of Functional Differential Equations. Appl. Math. Sci., Vol. 3, Springer-Verlag.

11) J. K. Hale [1984], Asymptotic behavior and dynamics in infinite dimensions. LCDS Report 84-28, Division of Applied Mathematics Brown University.

12) J. K. Hale, X.-B. Lin and G. Raugel [1985], Approximation of the in dynamical systems. In preparation.

13) J. K. Hale, L. T. Magalhães, and W. M. Oliva [1984], An Introduction to Infinite Dimensional Dynamical Systems - Geometric Theory. Appl. Math. Sci., Vol.47, Springer-Verlag.

14) J. K. Hale and C. Rocha [1985], Varying boundary conditions with large diffusion. LCDS Report 85-5, Div. Appl. Math., Brown University.

15) J. K. Hale and J. Scheurle [1985], Smoothness and bounded solutions of nonlinear evolutions equations. J. Differnetial Equations 56, 142-163.

16) A. Haraux [1984], Two remarks on dissipative hyperbolic systems. Preprint.

17) D. Henry [1981], Geometric Theory of Semilinear Parabolic Equations. Lect. Notes Math., Vol. 840, Springer-Verlag.

18) D. Henry [1983], Some infinite dimensional Morse-Smale systems defined by parabolic partial differential equations. Preprint, to appear in J. Differential Equations.

19) N. Krasovskii [1959], Stability of Motion (Russian), Moscow; English Translation, Stanford Univ. Press, 1963.

20) O. A. Ladyzenskaya [1972], A dynamical system generated by the Navier-Stokes equation. Trans. from Zapicki Nauka. Sem. Leningrad. Otdeleniya Mat. Inst. Steklov Acad. Nauk SSSR 27 (1972), 91-115.

21) O. A. Ladyzenskaya [1973], Soviet Phys. Dokl. 17, 647-649.

22) J. Mallet-Paret [1976], Negatively invariant sets of compact maps and an extension of a theorem of Cartwright. J. Differential Equations 22, 331-348

23) J. Mallet-Paret [1977], Generic periodic solutions of functional differential equations. J. Differential Equations 25(1977), 163--183.

24) J. Mallet-Paret, [1983] Morse decomposition and global continuation of periodic solutions of singulary perturbed delay equations. In Systems of Nonlinear Partial Differential Equations (Ed. J. Ball), Reidel, 1983.

25) J. Mallet-Paret and G. Sell [1985], in preparation.

26) R. Mañé [1981], On the dimension of the compact invariant sets of certain nonlinear maps. In Lecture Notes in Mathematics Vol. 898, 230-242, Springer-Verlag.

27) P. Massatt [1983], Attractivity properties of $\alpha$-contractions. J. Differen-Equations 48(1983), 326-333.

28) N. Minorsky [1962], Nonlinear Oscillations, van Nostrand, Princeton.

29) A. D. Mishkis [1951], Linear Differential Equations with Retarded Arguments (Russian) Izdat. "Nauka" Moscow.

30) D. Ruelle and F. Takens [1971], On the nature of turbulence. Comm. Math. Phys. 20(1971), 167-192; 23(1971), 343-344.

31) J. Shatah and W. Stauss [1985], Instability of nonlinear standing waves. Comm. Math. Phys.

32) P. E. Sobolevski [1965], Equations of parabolic type in a Banach space. Am. Math. Soc. Translations (2), 49(1965).

33) N. C. Sternberg [1985], Bound states of a nonlinear hyperbolic wave equation. Ph.D. Thesis, Div. Appl. Math., Brown University.

34) F. Takens [1981], Detecting strange attractors in turbulence. In Lect. Notes Math., Vol. 898, Springer-Verlag.

35) R. Temam [1983], Navier-Stokes Equations and Nonlinear Functional Analysis, CBMS-NSF Regional Conf. Series in Appl. Math. Vol. 41, SIAM, 1983.

36) V. Volterra [1909], Sulle equazioni integrodifferenziali della teorie dell' elasticità. Atti Reale Acad. Lincei 18(1909), 295.

37) V. Volterra [1928], Sur la théorie mathématique des phénomènes hérédi-taires. J. Math. Pures Appl. 7(1928), 249-298.

38) V. Volterra [1931], Théorie Mathématiques de la Lutte pour la Vie. Gauthier-Villars, Paris.

# AUTONOMOUS DIFFERENTIAL EQUATIONS FOR THE HÉNON MAP
# AND OTHER TWO-DIMENSIONAL DIFFEOMORPHISMS

Gottfried Mayer-Kress

Center for Nonlinear Studies

and

Theoretical Division

Los Alamos National Laboratory

Los Alamos, NM   87545

## ABSTRACT

For the study of bifurcations and chaotic behavior
of nonlinear dynamical systems, discrete mappings
turned out to be extremely useful. In order to
study the smooth evolution underlying two dimensional invertible mappings, we discuss the suspension of the Hénon map of the plane, the standard
map on the torus, and more general Cremona transformations of the plane.

## 1.   Introduction

For the description of dynamical long-time behavior of complex systems,
two kinds of models have been studied in the literature. The classical
models are based on ordinary differential equations, which means that a
continuous flow of time is considered. More recently, models with discrete
time-steps $\tau > 0$ have become popular, especially for the description of
erratic or chaotic behavior [1]. In this way, the frequencies of the
dynamical system above a maximal frequency $\omega_{max} = \frac{2\pi}{\tau}$ are neglected.
These models are based on ordinary difference equations and have been
successful in modelling different routes to turbulence [2]. In a general
sense it is clear that both of these approaches are equivalent in the
description of the different transitions that occur before chaotic behavior
sets in. However, little is known about how a given property of a discrete
model can be translated into a continuous model and vice versa. For

instance it is known for a unimodal map on the interval that the order of its maximum determines its universality class in the period-doubling route to chaos [3]; but the corresponding criterion for differential equations is not clear [4]. The main geometrical argument used in this work is based on the concept of Poincaré maps [5]. This is a geometrical method by which one can construct a discrete dynamical model from a recurrent continuous flow. Although it is now a standard method for the numerical analysis of chaotic flows to construct the corresponding Poincaré map, there are only very few examples [6] in which this Poincaré map can be given explicitly. On the other hand, similar difficulties are met if a differential equation should be associated with a given discrete map. Since discrete dynamical systems possess a natural high-frequency cut-off at $\omega_{max}$, the construction of a continuous-time suspension is by no means unique [13]. Therefore, in [13], [14] we constructed a continuous flow, which should come as close as possible to a damped and driven anharmonic oscillator. This is the natural analogue of an entire Cremona transformation because of the position independent damping factor. A different approach has been used by P. J. Channel [15].

In the second section we shall introduce the basic notions, then we shall show in the third section how a suspension is constructed (see Eq. (3.2)) from a given family of interpolating diffeomorphisms in the plane. We shall restrict ourselves to entire Cremona transformations in section 4 (Eq. (4.5)), and in section 5 we discuss suspensions of the standard map of the torus. Finally, we shall write down the explicit autonomous differential equations for this system in section 6 (Eq. (6.3)).

## 2. Basic notions

In this section we relate discrete time dynamics in the plane with continuous dynamics in the three dimensional space. Thus, let $T : \mathbb{R}^2 \to \mathbb{R}^2$ be a diffeomorphism in the plane, i.e., $T$ is a smooth, invertible map, with a smooth inverse $T^{-1}$. A discrete trajectory $(x_n, y_n)$ is created by $T$ via:

$$(x_{n+1}, y_{n+1}) = T(x_n, y_n) , \qquad (2.1)$$

where $n \in \mathbb{N}$ represents the discrete time-variable. Continuous-time dynamical systems, which are defined by differential equations, generate trajectories of a flow which is defined by:

<u>Def. 2.1</u>: Let $M_0 \subset \mathbb{R}^3$ be some smooth manifold, and let $\text{Diff}(M_0)$ be the space of all diffeomorphisms on $M_0$. Then a mapping $\phi : \mathbb{R} \to \text{Diff}(M_0)$, $t \to \phi_t$ is called a (continuous) <u>flow</u> on $M_0$ iff:

(i) $\phi_0 = \text{id}$,

(ii) $\phi_{s+t} = \phi_s \circ \phi_t$ for $s, t \varepsilon \mathbb{R}$ .

Here "id" is the identity mapping and "o" denotes composition of maps. In physical language, $\phi_t$ shifts some initial value $\underline{x}_0 \varepsilon M_0$ along its trajectory up to the point $\phi_t(\underline{x}_0) =: \underline{x}(t)$, which corresponds to a time-interval of length t. In his fundamental paper [7], Smale described how a continuous flow can be constructed as a suspension of some diffeomorphism. To this end we have (cf. [8]).

<u>Def. 2.2</u>: Let $M_0 \subset \mathbb{R}^3$ be a smooth manifold with closed subset $S \subset M_0$. Let $T : S \to S$ be a diffeomorphism on S and $\phi$ a flow on $M_0$. If there exist $\alpha, \beta \varepsilon \mathbb{R}^+$ such that:

$$(i) \quad \bigcup_{t \varepsilon (0,\alpha)} \phi_t(S) = M_0 \ ,$$

$$(ii) \quad \bigcup_{t \varepsilon (0,\beta)} \phi_t(S) =: S_\beta \subset M_0 \text{ open such that: } S_\beta \cap S = \phi \ ,$$

$$(iii) \quad \phi_\beta \big|_S = T \ ,$$

then $\phi$ is called a <u>constant-time suspension</u> of T.

The diffeomorphism T is then called the <u>Poincaré</u> map of the flow $\phi$.

Statement (iii) means that for the return time $\beta$, which determines the maximal frequency resolved by T, the restriction of the diffeomorphism $\phi_\beta \varepsilon \text{Diff}(M_0)$ to the subsurface S coincides with the diffeomorphism T on S. The construction of Smale [7] is based on the manifold $M_T$ which is defined by:

$$M_T := \{(x,y,u) \ \varepsilon \mathbb{R}^3 \ ; \ (x,y,u+1) \sim (T(x,y),u)\} \ , \tag{2.2}$$

where "$\sim$" means that the corresponding points are identified. The variable u corresponds to a "phase angle" and $M_T$ can be thought of as some "generalized Möbius-strip". The suspension $\tilde{\phi}$ on $M_T$ can now be defined as:

$$\tilde{\phi}_t(x,y,u) := (x,y,u+t) \ . \tag{2.3}$$

We can see that for a cross-section $S \subset M_T$, which is given by:

$$S := \{(x,y,u) \ \varepsilon \ M_T \ : \ u = 0\} , \tag{2.4}$$

the suspension $\tilde{\phi}$ reproduces T in the sense that: $\tilde{\phi}_1|_S = T$, i.e.,
$\tilde{\phi}_1(x,y,0) = (T(x,y),0)$. From the construction of the suspension
described above, it becomes clear that all of the dynamics of the system
is built into the structure of the manifold $M_T$ such that the explicit
structure of $\tilde{\phi}$ appears to become trivial.

In the following, we will use an approach which is physically more
intuitive. Instead of the manifold $M_T$, we consider the cylinder M defined
by:

$$M := \{(x,y,u) \ \varepsilon \ \mathbb{R}^3 \ ; \ (x,y,u+1) \sim (x,y,u)\} . \tag{2.5}$$

That means u is a cyclic variable which can be interpreted as a phase-
angle, e.g., of some periodic driving force. This interpretation conforms
to our motivation for this work, which is to model a periodically driven
damped anharmonic oscillator starting from a discrete two-dimensional map.

3. Construction of the suspension from a given family of interpolating
diffeomorphisms

3.1 Continuous suspension

In this section we suspend the diffeomorphism T of the plane (see
section 2). To this end, we introduce a family of interpolating diffeo-
morphisms $F_t \ \varepsilon \ \text{Diff}(\mathbb{R}^2)$. In sections 4 and 5 we shall construct explicit
examples for $F_t$.

Def. 3.1: Let $F_t : \mathbb{R}^2 \rightarrow \mathbb{R}^2$ be a diffeomorphism for all $t \ \varepsilon \ \mathbb{R}$ such that:

(i)  $F_o = \text{id}$ ,

(ii) $F_1 = T$ ,

(iii) $F_t = F_\varepsilon \circ T^n$ if $t = n + \varepsilon > 1$ , $n \ \varepsilon \ \mathbb{Z}$ ,

where we have used the abbreviation $T^n := T \circ T^{n-1}$, $T^o := \text{id}$. If $F_t$
depends smoothly on the time-parameter t, we call it an interpolating
diffeomorphism for T.

The requirement of smooth dependence of t implies that the Jacobian-determinant of $F_t$ also has to be a smooth function of t. This fact is important because diffeomorphisms $\hat{T}$ with negative Jacobian cannot be suspended and therefore cannot be considered as the Poincaré map of some continuous flow. Otherwise, there had to exist a $t^*\varepsilon(0,1)$ for which the Jacobian of $F_{t^*}$ would vanish. This is because of the fact that $F_o = \mathrm{id}$, which implies: $\det F_o = 1 > 0 > \det F_1 = \det \hat{T}$ and because of the continuity of $\det F_t$ as a function of t. This means that $F_{t^*}$ is not a diffeomorphism, which is in contradiction to the definition 3.1.

We now can construct a class of suspensions for a given diffeomorphism T in the plane by the following:

Proposition 3.1: Let M be the cylinder defined in (2.5) and let $\Sigma$ be the cross-section defined by:

$$\Sigma := \{(x,y,u) \ \varepsilon \ M \ , \ u = 0\} . \tag{3.1}$$

Let $F_t$ be an interpolating diffeomorphism for a given diffeomorphism T defined on $\Sigma$, where $t = n + \varepsilon \in \mathbb{R}$, $n \ \varepsilon \ \mathbb{Z}$, $\varepsilon \in [0,1)$. Then the mapping $\phi : \mathbb{R} \to \mathrm{Diff} \ M$, $t \mapsto \phi_t$ defined by:

$$\phi_t(x,y,u) := (F_{u+\varepsilon} \circ T^n \circ F_u^{-1}(x,y), u + \varepsilon) \tag{3.2}$$

is a constant-time suspension of T.

Remark: We identify the family of cross sections $\phi_t(\Sigma) \subset M$ with the plane $\mathbb{R}^2$.

Proof of prop. 3.1: First we have to show that $\phi$ describes a flow. Property (i) of def. 2.7 can be easily checked by insertion. For the part (ii), we have to distinguish between different cases.

1) case: $\phi_{n+\varepsilon} = \phi_n \circ \phi_\varepsilon$. From (3.2) we get

$$\phi_{n+\varepsilon}(x,y,u) = (F_{u+\varepsilon} \circ T^n \circ F_u^{-1}(x,y), u+\varepsilon) .$$

Now we insert the identity $\mathrm{id} = F_{u+\varepsilon}^{-1} \circ F_{u+\varepsilon}$ and obtain

$$\phi_{n+\varepsilon}(x,y,u) = (F_{u+\varepsilon} \circ T^n \circ F_{u+\varepsilon}^{-1} \circ F_u^{-1}(x,y), u+\varepsilon)$$

$$= \phi_n(F_{u+\varepsilon} \circ F_u^{-1}(x,y), u+\varepsilon)$$

$$= \phi_n \circ \phi_\varepsilon(x,y,u) \ .$$

Because of the asymmetrical role of n and $\varepsilon$ we show the

2) case: $\phi_{n+\varepsilon} = \phi_\varepsilon \circ \phi_n$. With the same argument as above, we get:

$$\phi_{n+\varepsilon}(x,y,u) = (F_{u+\varepsilon} \circ T^n \circ F_u^{-1}(x,y), u+\varepsilon)$$

$$= (F_{u+\varepsilon} \circ F_u^{-1} \circ F_u \circ T^n \circ F_u^{-1}(x,y), u+\varepsilon)$$

$$= \phi_\varepsilon(F_n \circ T^n \circ F_u^{-1}(x,y), u)$$

$$= \phi_\varepsilon \circ \phi_n(x,y,u) \ .$$

Furthermore, we have to show the

3) case: $\phi_{n+m} = \phi_n \circ \phi_m$. This is done by:

$$\phi_{n+m}(x,y,u) = (F_u \circ T_{a,b}^{n+m} \circ F_u^{-1}(x,y), u)$$

$$= (F_u \circ T_{a,b}^n \circ F_u^{-1} \circ F_u \circ T_{a,b}^m \circ F_u^{-1}(x,y), u)$$

$$= \phi_n(F_u \circ F_u^{-1}(x,y), u)$$

$$= \phi_n \circ \phi_m(x,y,u) \ .$$

Finally, we have to look at the 4) case: $\phi_{\varepsilon+\delta} = \phi_\varepsilon \circ \phi_\delta$. Here we have:

$$\phi_{\varepsilon+\delta}(x,y,u) = (F_{\varepsilon+\delta+u} \circ F_u^{-1}(x,y), u+\varepsilon+\delta)$$

$$= (F_{\varepsilon+\delta+u} \circ F_{u+\varepsilon}^{-1} \circ F_{u+\varepsilon} \circ F_u^{-1}(x,y), u+\varepsilon+\delta)$$

$$= \phi_\delta(F_{u+\varepsilon} \circ F_u^{-1}(x,y), u+\varepsilon)$$

$$= \phi_\delta \circ \phi_\varepsilon(x,y,u) \ .$$

For the case that one of the above sums, which appear as indices of $F_1$, becomes larger than one, we apply property (iii) of Def. 3.7. The conditions (i) and (ii) of Def. 2.2 are obviously satisfied by $\phi$, and property (iii) can be seen by noting that

$$F_o = F_o^{-1} = \text{id} .$$

Thus, we have arrived at a prescription which allows us to obtain a system of trajectories from T which depend continuously on the time-parameter t. In the next subsection we shall formulate conditions which will guarantee that these trajectories are also smooth. This smoothness is necessary for the trajectories to arise from differential equations.

## 3.2 Differentiable suspension

The next step in our procedure will be to examine under which conditions the time-derivative of the flow $\phi_t$ is a continuous function of t. From the definition in (3.2) and the linearity of the differentiation-operation we can see that, besides the trivial time-dependence of the u-component of $\phi_t$, only the function $F_{u+\varepsilon}$ will be affected. This means that we only have to require the differentiability of $F_t$ with respect to t. For $t = n+\varepsilon$ we obtain:

$$\frac{d\phi_t}{dt}(x,y,u) = (\frac{dF_{u+\varepsilon}}{d\varepsilon}(X,Y),1) , \tag{3.3}$$

where $\quad (X,Y) := T^n \circ F_u^{-1}(x,y) .$

Thus we have to make sure that $dF_{u+\varepsilon}/d\varepsilon$ is continuous for all $(X,Y)$. To make this condition explicit, we may choose as initial condition a point which lies in $\Sigma$. Then Eq. (3.3) will be replaced by

$$\frac{d\phi_t}{dt}(x,y,0) = (\frac{dF_\varepsilon}{d\varepsilon}(T^n(x,y)),1) . \tag{3.4}$$

Note that this differential equation is not autonomous, since it depends explicitly on the time variable $t = \varepsilon$ and on the initial conditions $(x,y) \varepsilon \Sigma$. Equation (3.4) represents a system of trajectories which are smooth in the open interval $(0,1)$ and, according to the definition $(3.2)$, also are smooth in every open interval $(n,n+1)$. We now have to formulate the condition that the system passes smoothly through the Poincaré sections at $t = n$. Thus, we have to consider the two one-sided limits

(i) $\lim\limits_{t \to 1_-} \dfrac{d\phi_t}{dt} (x,y,0) = \lim\limits_{\varepsilon \to 1_-} (\dfrac{dF_\varepsilon}{d\varepsilon}(x,y),1)$ ,

$$(3.5)$$

(ii) $\lim\limits_{t \to 1_+} \dfrac{d\phi_t}{dt}(x,y,0) = \lim\limits_{\varepsilon \to 0_+} (\dfrac{dF_\varepsilon}{d\varepsilon}(T(x,y)),1)$ .

The requirement that the limits coincide is a continuity condition for the derivative of the interpolating diffeomorphism $F_\varepsilon$.

## 4. Suspensions of entire Cremona transformations

In this section we will consider diffeomorphisms in the plane with a constant Jacobian. They can be written in the form:

$$T\binom{x}{y} = \begin{pmatrix} f_a(x)+y \\ -jx \end{pmatrix} ,$$

$$(4.1)$$

where $j > 0$ and $f_a(x)$ is a nonlinear function on the real line with a as the relevant parameter. Note that for the case $f_a(x) = 1 - ax^2$ the diffeomorphism T corresponds to the Hénon-system [9], which is well studied for its chaotic aspects. It is also among the simplest models which create a "horseshoe" in the plane (cf. [8]). The Jacobian of T is simply given by

$$\det J_T\binom{x}{y} = j ,$$

$$(4.2)$$

where

$$J_T\binom{x}{y} = \begin{pmatrix} f'(x) & 1 \\ -j & 0 \end{pmatrix} .$$

$$(4.3)$$

The interpolating family of diffeomorphisms $F_t$ for a suspension of T is given in the following proposition:

Proposition 4.1: Let $\xi$, $\eta$ $\varepsilon \, \mathscr{C}^1([0,1])$ such that $\xi(t)^2 + y(t)^2 > 0$ for all $t$ $\varepsilon$ $[0,1]$. Assume that the following boundary conditions are satisfied by $\xi$ and $\eta$.

(i)  $\xi(0) = \eta(1) = \dot{\xi}(0) = \dot{\xi}(1) = \dot{\eta}(1) = 0$ ,

(ii)  $\xi(1) = \eta(0) = 1$ ,                                              (4.4)

(iii)  $\dot{\eta}(0) = \ln j$ for $j > 0$ .

Furthermore, let $\zeta(t) = \dfrac{j^t}{\xi(t)^2+\eta(t)^2}$. Then we have an interpolating family

of diffeomorphisms for the entire Cremona transformation T of (4.1) given by:

$$F_t\begin{pmatrix}x_o \\ y_o\end{pmatrix} = \begin{pmatrix} \eta(t)\zeta(t)x_o+\xi(t)^2 f(x_o)+\xi(t)y_o \\[2ex] -\xi(t)\zeta(t)x_o+\xi(t)\eta(t)f(x_o)+\eta(t)y_o \end{pmatrix} .$$                (4.5)

The proof of this proposition can be found in [13], [14].

For an explicit realization we have chosen:

$$\xi(t) = (3-2t)t^2 \ , \quad \eta(t) = 1+\ell nj\ t-(2\ell nj+3)t^2+(2+\ell nj)t^3 \ .$$        (4.6)

The corresponding suspension of the Hénon-system can be visualized in Fig. 1. We have also produced a 16 mm film showing the dynamics of the folding processes generated by this suspension. In the same way as seen in Fig. 2, we have computed a series of Poincaré sections of the attractor for subsequent times. The dynamics are visually equivalent to the dynamics generated by the differential equation of a driven and damped pendulum like the Duffing oscillator [10]. In the movie one can clearly see how the fractal structure of the chaotic attractor arises. (The same technique has been previously used by J. P. Crutchfield, J. D. Farmer, N. Packard and R. Shaw [11] to illustrate the folding process in different chaotic flows.)

## 5. Suspension of the Standard Map

One of the most popular mappings studied in connection with bifurcations and chaos of nonlinear dynamical systems is the Chirikov standard map $C(q,p)$, which is defined by (see e.g. [16]):

$$C(q,p) = (q+p+f(q),p+f(q)) \bmod 1 , \qquad (5.1)$$

$$\text{where} \quad f(q) = -\frac{k}{2\pi} \sin(2\pi q) .$$

Since it is defined on the two-torus $T^2$ rather than on the plane, and since, under the action of the map C, the torus $T^2$ is "twisted"; it cannot be smoothly transformed into its image under **C,** and therefore the previous method cannot be applied directly. The reason for that is that - although the standard map itself leaves the torus invariant - its interpolating function for intermediate times does not. Thus, the suspension of the standard map has to be defined on a larger manifold. Instead of the torus $T^2$, which can be identified with a unit square of the plane with periodic boundary conditions, we consider the covering of the torus. By this we mean that we distinguish between points on the torus $T^2$, which differ by an integer, or are contained in different squares in the corresponding planar representation. Therefore, we can introduce the interpolating diffeomorphisms $C_t$ of the standard map as functions of the plane to itself. For integer values of t = n the flow will be identical with the iterated standard map $C^n$, and therefore invertible on the torus $T^2$ itself. In order to construct the suspension of the standard map explicitly, we write it in the following form:

Let $E(q,p) = (q,p-q) =: (q,r)$ a linear coordinate transformation of the plane and $E^{-1}(q,r) = (q,r+q)$ its inverse. Then we get a new mapping $\tilde{C}(q,r)$ as

$$\tilde{C}(q,r) = E \circ C \circ E^{-1}(q,r) = (F(q)+r,-q) , \qquad (5.2)$$

where $F(q) = f(q) + 2q$ and $f(q)$ from Eq. **(4.1)** .

With this transformation we have the lifted standard map in our canonical form of an entire Cremona transformation (see Eq. (4.1) with j = 1). The more general dissipative standard map (with j < 1) can be obtained by application of the inverse coordinate transform:

$$C(q,p) = E^{-1} \circ \tilde{C} \circ E(q,p)$$

$$= (F(q) + r, F(q) + r - jq) \tag{5.3}$$

$$= (f(q) + p + q, f(q) + p + (1-j)q) .$$

In the numerical calculations, we consider the intersection of the image of the plane (or covering manifold) under $C_t$ with the unit square (or the torus $T^2$). We can confine ourselves to times $0 < t < 1$ since, for other values of t, we start with the image of $T^2$ under an integer number of applications of the standard map C. Thus, we also can restrict ourselves to small neighborhoods of the torus for the domain of $C_t$. In Fig. 3 we see a sequence of images of horizontal lines under $C_t$ for different values of t.

## 6. Autonomous differential equations from the suspension of the entire Cremona transformations

In the previous section we constructed a smooth suspension for the diffeomorphism (4.1). Thus, we can consider Eq. (4.5) as a solution of the differential-equation (4.8). However, Eq. (4.8) represents a non-autonomous system since it depends on the initial values $x_0$, $y_0$ and not on the actual position at time t, which is given by:

$$\begin{pmatrix} x(t) \\ y(t) \\ z(t) \end{pmatrix} = \begin{pmatrix} F_t \begin{pmatrix} x_0 \\ y_0 \end{pmatrix} \\ \omega t + t_0 \end{pmatrix} , \tag{6.1}$$

where in our case $\omega = 1$ and $t_0 = 0$. In order to transform Eq. (4.8) into an autonomous differential equation, we first have to express the initial values $x_0$, $y_0$ as a function of $x(t)$, $y(t)$. After straight-forward but somewhat lengthy calculations we obtain:

$$x_0(x(t), y(t)) = j^{-t}(\eta(t)x(t) - \xi(t)y(t)) ,$$

$$y_0(x(t), y(t)) = \frac{\xi(t)x(t) + \eta(t)y(t)}{\xi(t)^2 + \eta(t)^2} - \xi(t)f(x_0) . \tag{6.2}$$

When we now insert the initial conditions of Eq. (6.2) into Eq. (4.8), we can express $\dot{x}(t)$, $\dot{y}(t)$ and $\dot{z}(t)$ as a function of $x(t)$, $y(t)$, and $z(t)$, and thus have made the differential equations autonomous. They are given by

$$\dot{x}(t) = h(\xi(t),\eta(t))x(t)+g(\xi(t),\eta(t))y(t)+\xi(t)\dot{\xi}(t)f(x_0(x(t),y(t))) \, ,$$

$$\dot{y}(t) = g(\eta(t),\xi(t))x(t)+h(\eta(t),\xi(t))y(t)+\dot{\xi}(t)\eta(t)f(x_0(x(t),y(t))), \quad (6.3)$$

$$\dot{z}(t) = \omega \, ,$$

where $x_0(x(t),y(t))$ is taken from Eq. (6.2), and the functions g and h are given by

$$g(\xi,\eta) = \frac{1}{\xi^2+\eta^2} \{\dot{\xi}\eta-\xi\dot{\eta}-\xi\eta(\ell nj-2 \, \frac{\xi\dot{\xi}+\eta\dot{\eta}}{\xi^2+\eta^2})\} \, ,$$

$$\hspace{10cm} (6.4)$$

$$h(\xi,\eta) = \frac{1}{\xi^2+\eta^2} \{(\xi\dot{\xi}+\dot{\eta}\eta) \, \frac{\xi^2-\eta^2}{\xi^2+\eta^2} + \eta^2\ell nj\} \, .$$

Here we have omitted the arguments where no confusion was possible. We see that the only nonlinearity comes indeed from the function f of Eq. (4.1). The main complexity, however, is contained in the t- (or z-) dependence of the coefficient functions g and h. It would be interesting to see whether it is possible to express g and h as functions of $x(t)$ and $y(t)$ alone. Finally, we will show that this autonomous O.D.E. (6.3) in fact represents a driven and damped oscillator with constant dissipation. For this purpose we evaluate the divergence of the vector field in Eq. (6.3). It is given by

$$\text{div}\begin{pmatrix}\dot{x}\\\dot{y}\\\dot{z}\end{pmatrix} = \frac{\partial\dot{x}}{\partial x} + \frac{\partial\dot{y}}{\partial y} + \frac{\partial\dot{z}}{\partial z}$$

$$= h(\xi,\eta)+\dot{\xi}\xi f'(x_o)\cdot j^{-t}\cdot\eta+h(\eta,\xi)$$

$$+ \dot{\xi}\eta f'(x_o)(-\xi j^{-t})$$

$$= h(\xi,\eta)+h(\eta,\xi) \ .$$

When we now insert the explicit form of $h(\xi,\eta)$ from Eq. (6.4) we get:

$$\text{div}\begin{pmatrix}\dot{x}\\\dot{y}\\\dot{z}\end{pmatrix} = \frac{1}{\xi^2+\eta^2} \ \{\frac{(\dot{\xi}\xi+\dot{\eta}\eta)}{\xi^2+\eta^2} \ (\xi^2-\eta^2+\eta^2-\xi^2)$$

$$+ \ell n j(\eta^2+\xi^2)\}$$

$$= \ell n j \ .$$

This is exactly the "damping constant" which we would expect from Eqs. (4.2) and (4.8).

## 7. Conclusion

We have constructed an explicit solution for a class of chaotic autonomous ordinary differential equations, which is expressed by analytical functions and iterates of a diffeomorphism. The method we have used is based on the notion of smooth suspensions. We have also made a concrete connection between diffeomorphisms in the plane and continuous flows in three-dimensional space. Our emphasis was laid on maps with constant Jacobian and, correspondingly flows with a position-independent divergence. These cases correspond to a constant damping term in the equations for a driven and damped oscillator. Thus, we have constructed a way to study chaotic dynamics of continuous systems by calculating analytical maps instead of numerically integrating the differential equations.

306

References

[1] P. Collet, J.-P. Eckmann, "Iterated maps on the interval as dynamical systems" (Birkhäuser, Boston, 1980).

[2] J.-P. Eckmann, Rev. Mod. Phys. 53, 643 (1981);
D. Ruelle, F. Takens, Commun. Math. Phys. 20, 167 (1971);
Y. Pomeau, P. Manneville, Commun. Math. Phys. 77, 789 (1980).

[3] M. Feigenbaum, J. Stat. Phys. 19, 25 (1978).

[4] S. Grossmann, private communication.

[5] See, e.g., H. Haken, "Advanced Synergetics" (Springer Berlin, 1983).

[6] O. L. González, O. Pira, Phys. Rev. Lett. 50, 870 (1983).

[7] S. Smale, Bull. Amer. Math. Soc. 73, 747 (1967).

[8] J. D. Farmer, J. P. Crutchfield, H. Froehling, N. Packard, R. Shaw, "Power spectra and mixing properties of strange attractors," Proc. of the 1979 Conf. on Nonlin. Dyn. (ed. R. H. G. Helleman), Ann. N.Y. Acad. Sci. 375, 353 (1980).

[9] M. Hénon, Commun. Math. Phys. 50, 69 (1976).

[10] H. Bunz, private communication.

[11] J. P. Crutchfield, J. D. Farmer, N. Packard, and R. Shaw, "Mixing properties of chaotic attractors," 16 mm film.

[12] O. E. Rössler, Phys. Lett. 57A, 196 (1976).

[13] G. Mayer-Kress, "Zur Persistenz von Chaos und Ordnung in nichtlinearen dynamischen Systemen," Ph.D. thesis, Universität Stuttgart, 1984.

[14] G. Mayer-Kress and H. Haken, "An explicit construction of a class of suspensions and autonomous differential equations for diffeomorphisms in the plane," Los Alamos preprint, 1984.

[15] P. J. Channel, "Explicit suspensions of diffeomorphisms - an inverse problem in classical dynamics," J. Math. Phys. 24, 823 (1983).

[16] B. V. Chirikov, Phys. Rep. 52, 269 (1979).

Figure Captions

Fig. 1.  Projection of the solution $\begin{pmatrix} x(t) \\ y(t) \\ z(t) \end{pmatrix}$ of Eq. (6.1) for the

differential equation (6.3) in a chaotic parameter regime of
the Hénon-system (4.1) (a = 2.1, j = 0.3) onto the $(x_1, x_2)$-
plane.

Where:  $x_1(t) = (x(t)+1) \cos 2\pi t$

$x_2(t) = (x(t)+1) \sin 2\pi t.$

Note the similarity to the simple Rössler-attractor [12].  The
lines A,B,...,F indicate Poincaré sections at successive
angles.

Fig. 2.  Poincaré cross sections of the suspension of Fig. 1 at
successive angles denoted by A,B,...,F.  In order to illustrate
the folding-, stretching- and construction-process, we also
have computed the trajectories through an ellipse in a
neighborhood of the attractor.

Fig. 3.  Successive images of horizontal lines of covering of the two-
torus $T^2$ under the interpolating diffeomorphisms $C_t$ section 5
for $t \in \{0, \frac{1}{2}, 1, 1\frac{1}{2}, 2, 2\frac{1}{2}, 3\}$.

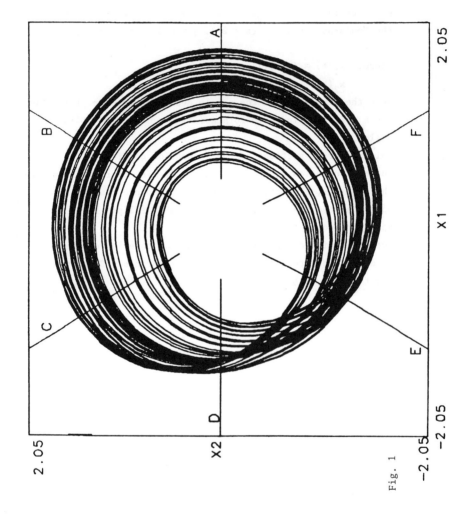

Fig. 1

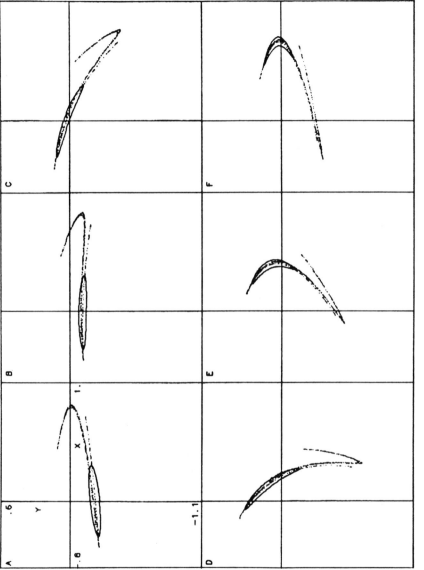

Fig. 2

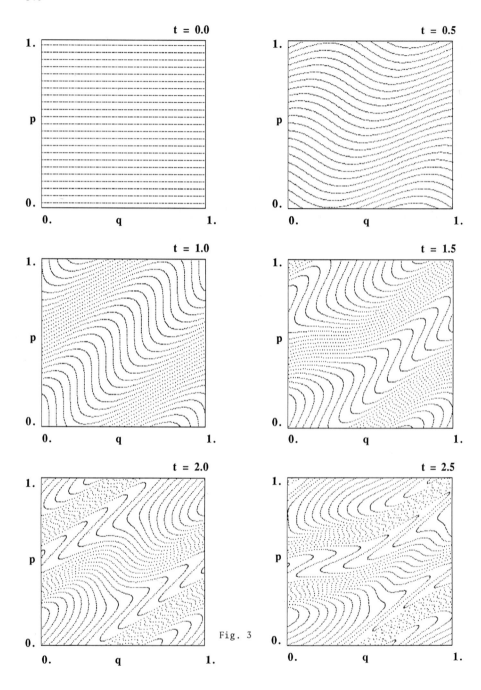

Fig. 3

Bifurcation and the Integration of Nonlinear Ordinary and Partial Differential
Equations

by

M.S. Berger*

INTRODUCTION

I would like to begin this article with a question:

Will the recent advances made in the nonlinear problems of dynamics and mathematical physics make a fundamental change in the presentation of nonlinear aspects of science in academic texts?

I believe the answer to this question is in the affirmative. I would like to point out here one reason for my optimism. Perhaps I should describe the words in my title. Bifurcation I view as a basic (possibly) infinite dimensional nonlinear phenomenon. It was formally initiated by Poincaré in the 1880's. The phenomenon arises for a nonlinear problem in which the standard methods for the inverse function theorem or computationally for the Newton method break down. Thus, the formal linearization for a nonlinear problem needs more careful scrutiny. Nonlinear instability may result and moreover the mathematical approaches to understanding of the problem remained a clouded mystery until the late 1960's.

On the other hand, the integration of nonlinear differential equations in closed form has a long history dating back to the beginnings of calculus. The entire notion became known as "integrability by quadrature." Here in this article I am going to study a cause for the breakdown in the notion of integrating by quadrature. Indeed it is known that many relatively simple nonlinear ordinary differential equations cannot be integrated by quadrature. The classic example is the Riccati equation (as discussed in the book <u>Theory of Bessel Functions</u>, by G.N. Watson, Chapter 4).

My goal in the present article is to discuss an extension of integrability by quadrature (i.e. closed form solutions) that incorporates bifurcation as a fundamental ingredient.

* Research partially supported by an NSF grant and an AFOSR grant.

### 1. A Key Problem of Nonlinear Dynamics

As was mentioned in the introduction this article is devoted to the question:

Can integrability be found when bifurcation is present?

This question by its very nature requires a mathematical analysis. First, of the notion of integrability and secondly, of the bifurcation process.

Intuitively we are dealing with two extreme situations in nonlinear dynamics. First, there is the notion of a completely integrable dynamical system. Such systems are famous in classical dynamics treatises, for example, the book of E.T. Whittaker, and the names of Hamilton-Jacobi and Liouville are called to mind in this connection. For partial differential equations notions concerning the KdV equation and nonlinear Schrödinger equation in one space dimension are well known to be integrable by the inverse scattering method.

On the other hand, many nonlinear dynamical systems exhibit the reverse of integrability, namely "chaotic" behavior; by which we mean sensitivity to perturbations of the initial conditions, as well as unpredictable asymptotic behavior. Our goal in this work is to find a case intermediate between these two extremes that incorporates bifurcation phenomenon as an intrinsic part of the integrability idea.

The significance of our extension is the realization that the integrability idea is a global one whereas bifurcation and chaotic behavior is formulated from a local point of view. Thus we must attempt to understand bifurcation phenomena more globally in order to address our fundamental key problem. This will be one of our objects in this paper.

**2.** Transformation Theories for Hamiltonian Systems of Finite and
Infinite Dimension

Part 1: Finite Dimensional Systems

The usual Hamiltonian equations of a finite dimensional system with **N**
degrees of freedom can be written:

$$\frac{dp_i}{dt} = \frac{-\partial H}{\partial q_i} \, ,$$

$$\qquad (i = 1, 2, \ldots n) \qquad (1.)$$

$$\frac{dq_i}{dt} = \frac{+\partial H}{\partial p_i} \, .$$

The integrability idea as discussed in Liouville's theorem and the Ham-
iltonian-Jacobi separation of variables idea require N conservation laws
($F_i$ = constant, i = (1,2, . . . )) that are independent and the respective
Poisson brackets are in involution. One attempts to introduce angle-ac-
tion variables in such a way that certain canonical transformations conserv-
ing the Hamiltonian structure are preserved. More formally we attempt to find
a canonical transformation

$$(p,q) \rightarrow (F,\Phi)$$

such that **(1.)** is transformed into

$$\frac{dF_i}{dt} = 0$$

$$\qquad (2.)$$

$$\frac{dI_i}{dt} = c_i(F)$$

An important fact to notice is that on the energy surface

$$M = \{x \big| F_i(x) = c_i \ (i = 1, 2, \ldots N)\}$$

the system (1.) becomes linear after the desired canonical transformation is
performed to reduce (1.) to (2.)

314

Part 2:  Infinite Dimensional Systems

In considering nonlinear partial differential equations with a Hamiltonian structure it has been customary to consider nonlinear evolution equations of the form

$$\frac{\partial u}{\partial t} = \Sigma(u) \qquad\qquad (3.)$$

supplemented by boundary conditions on the spacial components of u.  Then one attempts to study the initial value problem for (3.) by making a formal change of coordinates to simplify (3.), hopefully to a problem with linear structure.

This was accomplished for the Burgers equation in the 1930's and has been extended greatly for nonlinear evolution involving one spatial dimension via the inverse scattering method.  This method requires an infinite number of independent conservation laws and instead of focussing on the Poisson Bracket formulation one rewrites the nonlinear evolution equation in terms of the commutator formalism of P. Lax  so that the nonlinear evolution equation has the form

$$\frac{\partial u}{\partial t} = [L,B], \text{ where } [L,B] \text{ denotes the commutator of } L \text{ and } B.$$

Here L and B represent differential operators.  Here the following diagram illustrates the inverse scattering method with scattering transformation S and appropriate $S^{-1}$ inverse scattering data.

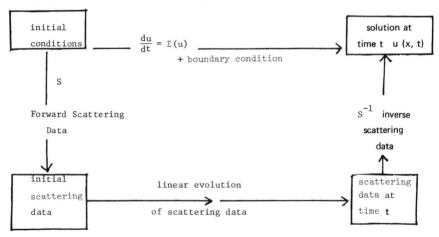

Here an amazing linearization occurs since the scattering data of the linear
evolution is associated with the spectrum of the differential equation

$$Lu + \lambda u = 0 \quad + \text{ null boundary conditions at infinity.} \qquad (4.)$$

The discrete part of this spectrum of L gives rise to solitons and multi-
solitons. This discovery is remarkable and has caused tremendous excitement in
present day mathematical research.

Research credit for such discoveries belongs to many people:  Kruskal and
Zabusky, McKean, Novikov, Lax, among many others.

It is rather remarkable that the number of equations of mathematical physics
that can be studied by this inverse scattering method is rather large.  In fact,
there are at least seventeen examples of such equations.  However, they all
involve spatial dependence in one variable (with the notable exception of the
KP equation).  All these equations cannot be perturbed since the conservation
laws necessary for their integration are destroyed.  Moreover, it is extremely
curious that the nonlinear equations involved can be completely studied by a
nonlinear problem, namely, the spectral properties of the operator  L .  This
led me to the conclusion that there must be a whole new theory of integrability
for higher dimensional nonlinear evolution equations since these equations of
mathematical physics generally do not have an infinite number of conservation
laws.  The next step in my thoughts was the realization that bifurcation was
excluded from the ideas of integrability in both the ideas of Section 2.  In
the remaining sections I shall outline my new integrability idea in fairly
loose terms.  The interested reader can find more information in some of the
articles listed in the bibliography at the end of my article.  Because the
audience for this article is not primarily composed of mathematicians, I have
relegated the few proofs contained here to a small appendix.  Once again,
many of the other proofs will be found in the articles of the bibliography.
However, it should be added that the glory of mathematics is in its proofs
and its ideas so that proofs clarifying the concepts involved are absolutely
essential to keep a researcher from the pitfalls of error.

### 3. Reasons for Research on Integrability

The success of the transformation theories mentioned in Section **2** leads
one to expect further answers for more complicated nonlinear systems of differential equations. Pertinent research questions are:

(1)  What is behavior under perturbation?

Change of parameter?

Large driving term?

(2)  How does one deal with Bifurcation and instability phenomena?

(3)  Is it possible to extend integrability of partial differential equations of two or more space dimensions?

(4)  In the above three questions there is a need for a synthesis of the
presence of these problems with twentieth century nonlinear analytic
ideas.

As mentioned above, the problem of integrability is by its very nature a
mathematical one since physical common sense does not aid in the strict analytical
processes involved. One utilizes the special functions that are famous in
mathematical physics, all the applied mathematics built up over the last few
centuries, and to do this one adds the modern striking new elements of mathematical research. Taken together, all these objects form an immensely powerful
research engine for new truths. Just as in the days of Maxwell and Faraday,
it is necessary for experimental and theoretical work to march hand in hand.
Taken together with the modern high-speed digital computer, the ideas of classical
integrability can surely be greatly extended.

What we shall accomplish in the sequel is to construct a new systematic procedure for integrating nonlinear differential equations explicitly in cases not previously handled by older techniques of global linearization. Fundamental in our idea is that the new high speed digital computer can be used in cases where bifurcation occurs and linearization techniques fail. These new ideas are global and so it is necessary to describe the mathematical concepts associated with them. It turns out that our systematic procedure for integrability has two distinct parts. One part analytic and one part geometric. The idea is that we distinguish new invariants for integrability. These invariants are the bifurcation points of a certain nonlinear problem, here called singular points. Differential typology shows that to these invariants must be added the bifurcation values of a given nonlinear problem. For problems that can be integrated in our sense, the two geometric sets have a very simple structure and the essence of the integrability problem I envision is to determine the structure of these geometric sets.

Thus the analytical part of the integrability problem in my approach consists of four steps. This procedure is systematic and can be used for a wide variety of different problems (see Section 4).

Step 1: Reduction to a finite-dimensional problem.

Step 2: Explicit cartesian representation for the singular bifurcation points of A.

Step 3: Explicit cartesian representation for the singular bifurcation values of A.

Step 4: Coerciveness estimates for the associated nonlinear problem.

The second part is geometric, in which we construct the changes of coordinates that reduce a nonlinear problem to its simplest (see Section 4) form.

Step 1: We begin by successively changing coordinates so that a given nonlinear problem is "layered". This means simply that the reduction to a finite dimensional problem is carried out explicitly by mappings.

Step 2: We transform the bifurcation points of a given nonlinear problem into the bifurcation points of a canonical nonlinear problem that we label C.

Step 3: We transform the bifurcation values of a given nonlinear problem into those of the canonical problem C.

Step 4: Based on these three coordinate changes we transform the entire nonlinear problem into the canonical global normal form C.

### 4. Mathematical Statement of the Problem

Our goal is to study the explicit solutions of the equation

$$A_\lambda u = g \ . \tag{5.}$$

Here $A_\lambda$ is a nonlinear differential operator generally depending on a parameter $\lambda$ and $g$ is an inhomogenous term (or forcing function) that varies over a large class of functions. This equation (5.) differs substantially from the integrable systems discussed in Section II. First, because it depends explicitly on parameter $\lambda$, and secondly, because of the inhomogenous term $g$, on the right hand side of (5.). We also notice that $A$ is a mapping between function spaces that will correspond to a continuous, in fact differentiable mapping A, acting between function spaces $X$ and $Y$. Thus, I put functional analysis directly into the formulation of the problem to be considered.

Instead of initial conditions I shall study boundary value problems associated with the equation (5.). This means that I change the function spaces $X$ and $Y$ in accord with the boundary conditions associated with the equation (5.). For example, Poincaré has emphasized the role of periodic solutions in dealing with the intial value problems for finite dimensional Hamiltonian systems. Recently, Feigenbaum has emphasized the fact that in chaotic problems periodicity instead of initial value problems may be a crucial ingredient for understanding.

My idea is to apply a new transformation theory to equation (5.). The idea of the transformation theory for (5) is to make two changes of variables, $u = Wv$ and $g = Tf$, where $W$ and $T$ are global changes of coordinates, so that the equation (5) takes the form

$$C_\lambda v = f \ . \tag{6.}$$

Here $A_\lambda = WC_\lambda T^{-1}$ and the equation (6) has a simple canonical form. In pictures we have the following diagram

$$\tag{7.}$$

Our idea is to find invariants for this new transformation theory. Indeed the notion of conservation laws and Poisson Brackets do not seem to be relevant to this diagram provided W and T are not inverses of each other. (Of course if they are inverses the previous transformation theory of Section II applies and the operator C becomes a linear operator.) In the present situation our definition represents a nonlinear extension of the usual transformation theory diagram. Our idea is to substitute differential topology and its invariants for the conservation laws. In fact the diagram (7.) occurs in a local finite dimensional analog in the theory of singular points and singular maps in differential topology. In analysis singular points are called bifurcation points and this is the desired connection that we need for our extension. Differential topology adds a new invariant namely bifurcation values (that is the image of bifurcation points) under the given mapping A. In addition to all this differential topology gives a classification of singular points and to types invariant under the diagram (7.). This classification reduces the problem of bifurcation points to the study of local normal forms for mappings and in addition studies the stability of the associated singular points under perturbation.

Of course differential topology does not solve the entire analytical problem because the studies until now have been restricted to local finite dimensional ones and the problem at hand is global and infinite dimensional. These last two facts are the glory of nonlinear analysis. The first global property that we shall need is the various aspects of compactness "properness", which justify the choices of our function spaces X and Y. Moreover, we must make a deeper study of nonlinear differential operators. This properness property is discussed in my text Nonlinearity and Functional Analysis, Academic Press (1977).

## 5. Discussion of Examples

Before proceeding with the actual examples to illustrate our theory perhaps it is wise to repeat the goals we have in mind. These goals are six in number, and can be listed as follows:

(1)   Independence of space dimensions.

(2)   Stability under perturbation.

(3)   Inclusion of examples not integrable by quadrature.

(4)   Inclusion of parabolic nonlinear evolution equations as well as elliptic boundary value problems.

(5)   To discover new simple structures in equations of mathematical physics.

(6)   To create a systematic procedure to test the integrability of a large class of differential equations, and moreover, for equations that pass this test to find new explicit methods of solution that utilize the modern high speed digital computer.

(7)   Of course, our key idea is to include bifurcation phenomena in these examples. Thus, in general we do not consider initial value problems, but rather boundary value problems of Dirichlet or periodic type. This enables us to study multiple solutions for time-dependent nonlinear evolution equations in a comprehensive way and as well by studying periodicity to avoid the chaotic phenomena in initial value problems for a large family of evolution equations.

In the sequel the following global normal forms are important:

(i)    the Whitney fold  $(t,x) \rightarrow (t^2,x)$

(ii)   the Whitney cusp  $(t,\lambda,x) \rightarrow (t^3-\lambda t,\lambda,x)$.

## 6. Examples from Ordinary Differential Equations

The simplest example of an ordinary differential equation not integrable by quadrature is the Riccati equation

$$\frac{dy}{dt} + y^2 = f(t) . \tag{8.}$$

To study this equation by our methods we focus on the left-hand operator

$$A_y = \frac{dy}{dt} + y^2 , \tag{9.}$$

and instead of considering the initial value problem we focus on T-periodic boundary conditions

$$y(0) = y(T). \tag{10.}$$

The singular points of the operator $A$ turn out to be a hyperplane of co-dimension one in the Sobolev space $W_{1,2}(0,T)$. The singular values of the operator $A$ turn out to be the boundary of an infinite dimensional convex set in the Sobolev space. The associated operator $A$ is a proper map. In fact all the steps outlined in Section 3, in our systematic procedure for integrability can be carried out explicitly. This will be described in the papers of the bibliography. Here $X = L_2(0,T)$ and $Y = L_2(0,T)$ subject to (10.).

The final result is a global normal form for the operator $A$ and explicit coordinate transformations to achieve this normal form. The result is as follows:

Theorem: The mapping $A$ defined by equation (9.) together with the periodic boundary conditions (10.) is $C^1$-equivalent to a global Whitney fold by explicit coordinate changes. (See bottom of Section 7 and the definition of Section 8.)

The next simplest example of a system not integrable by quadrature is the Abel equation which can be written

$$\frac{dy}{dt} + P_3(y,\lambda) = f(t) . \tag{11.}$$

To set up this problem in our context we define an operator $A_\lambda$, as follows:

$$A_\lambda y = \frac{dy}{dt} + \lambda y - y^3, \quad y(0) = y(T). \tag{12.}$$

This operator once again has a large bifurcation set, and in fact this set consists of points all of which are folds and cusps. What is interesting here is that this operator is integrable in our new extended sense. In fact we prove

Theorem: $A_\lambda$ (defined by equation (12.)), is $C^1$—equivalent to a global Whitney cusp, where the Hilbert spaces X and Y are the same as above.

We can proceed even further with this work, and study the equation

$$\frac{dy}{dt} + P_N(y,\lambda) = f(t) \tag{13.}$$

together with the T-periodic conditions (10.). Once again we define the operator $A_\lambda$ to be

$$A_\lambda y = \frac{dy}{dt} + P_N(y,\lambda) \ , \tag{14.}$$

as a mapping between the Sobolev space $W_{1,2}(0,T)$ into $L_2(0,T)$, together with the periodic boundary conditions (10.). This mapping has the linearization

$$A_\lambda'(y)v = \frac{dv}{dt} + P_N'(y)v \ ,$$

$$v(0) = v(T) \ . \tag{15.}$$

This mapping (as will be proved in the last section of this article) has singular points exactly when

$$\int_0^T P_N'(y(t))dt = 0 \ . \tag{16.}$$

## 7. Examples from Partial Differential Equations

Here our goal is to create an integrability theory that is independent of space dimension. Moreover we shall follow the procedure outlined in Section 3. We consider the Euclidean Log Cosh Gordon Equation.

We begin with an arbitrary bounded domain $\Omega \subset \mathbb{R}^N$, and we consider the equation

$$A_\alpha u = f . \tag{17.}$$

Here $A_\alpha u = \Delta u + \alpha u + \beta \log \cosh u$ is supplemented by the Dirichlet boundary condition $u|_{\partial \Omega} = 0$. Here the constants $\alpha, \beta$ are positive and restricted by the lowest two eigenvalues $\lambda_1$, and $\lambda_2$ of the Laplace operator $\Delta$

$$0 < \alpha - \beta < \lambda_1 < \alpha + \beta < \lambda_2 . \tag{18.}$$

We then consider the nonlinear Dirichlet problem

$$A_\alpha u = f, \quad f \in L_2(\Omega) ,$$

$$u\big|_{\partial \Omega} = 0 . \tag{19.}$$

Theorem: The equation (19.) has either 0, 1, or 2 solutions depending on whether the size of the projection of f on Ker($\Delta + \lambda_1$) is less than, equal to or greater than a certain computable critical number.

To understand this result it is useful to investigate its complete integrability and, in fact we can prove:

Theorem: The operator $A_\alpha$ is completely integrable as a mapping between the Holder space $C^{2,\infty}(\overline{\Omega})$ and $C^{0,\infty}(\overline{\Omega})$ $(0 < \alpha < 1)$. In fact, $A_\alpha$ is $C^1$ conjugate to the diagonal mapping ("the global Whitney fold map")

$$(t,w) \rightarrow (t^2, w) , \tag{20.}$$

where a general element u of $C^{2,\infty}(\overline{\Omega})$ is written $u = t u_1 + w$ with $u_1$ a normalized positive eigenvector of $\Delta$ associated with $\lambda_1$.

**8.** Special Stability Properties of the New Integrability Method

The methods that we have been advocating in this article have a very special feature stemming from the mathematical structures from the bifurcation that we utilize. Conventional ideas on complete integrability and inverse scattering techniques break down when the integrable equations involved are perturbed. However, the methods we are advocating are stable under perturbation, provided the bifurcation points involved are folds or cusps. The global normal forms involved will not be destroyed under perturbation, but rather the changes of coordinates defining the global normal forms will be slightly perturbed.

To illustrate this we give a few definitions of these special stable singular points, folds, and cusps.

Definition: Let $A$ be a $C^k (k \geq 2)$ map germ at $\bar{u} \in E_1$. Then $A$ is a <u>fold</u> if:

(0) $A$ is Fredholm at $\bar{u}$ with index 0;

(1) dim ker $DA(\bar{u}) = 1$ (and therefore range $DA(\bar{u})$ has codimension one); and

(2) for some (and hence for any) nonzero element $e_0 \in \ker DA(\bar{u})$,

$$D^2 A(\bar{u})(e_0, e_0) \notin \text{range } DA(\bar{u}).$$

For a map $A$ a point $\bar{u}$ in its domain is called a <u>fold point</u> if the germ of A at $\bar{u}$ is a fold.

Definition: Let $A$ be a $C^k (k \geq 2)$ map germ at $\bar{u} \in E_1$. Then $A$ is a <u>precusp</u> if:

(0) $A$ is Fredholm with index 0;

(1) dim ker $DA(\bar{u}) = 1$ (and therefore range $DA(\bar{u})$ has codimension one);

(2) for some (and hence for any) nonzero $e_0 \in \ker DA(\bar{u})$,

$$D^2 A(\bar{u})(e_0, e_0) \in \text{range } DA(\bar{u}); \text{ and}$$

(3) for some $\omega \in E_1$,

$$D^2 A(\overline{u})(e_0, \omega) \notin \text{range } DA(\overline{u}).$$

If $k \geq 3$ and $A$ is a precusp satisfying

$$(4) \quad D^2(A|SA)(\overline{u})(e_0, e_0) \notin \text{range } D(A|SA)(\overline{u})$$

(where $SA$ is the singular set or critical set of A), then $A$ is a <u>cusp</u>.

If $k \geq 3$ condition (4) can be changed to

$$(\tilde{4}) \quad D^3 A(\overline{u})(e_0, e_0, e_0) - 3D^2 A(\overline{u})(e_0, (DA(\overline{u}))^{-1}(D^2 A(\overline{u})(e_0, e_0))) \notin \text{range } DA(u).$$

Of our examples, certainly the Riccati equation of Section 6 and the Abel equation have a remarkable stability property. In addition to that the example of Section 7 which we called the Euclidean Log Cosh Gordon Equation is also stable since its singularities consist entirely of folds. More explicitly we have the following theorem:

Theorem: The <u>global</u> <u>normal</u> <u>form</u> (20.) <u>is</u> "<u>stable</u>" <u>for</u> A <u>in the</u> <u>sense</u> <u>that</u> <u>if</u> <u>the</u> <u>log</u> <u>cosh</u> <u>Gordon</u> <u>operator</u>

$$Au = \Delta u + \alpha u + \beta \log \cosh u, \quad u|_{\partial\Omega} = 0$$

<u>is</u> <u>perturbed</u> <u>to</u> $\tilde{A}(u) = \Delta u + f(r), u|_{\partial\Omega} = 0$ <u>with</u>

$$||f(u) - (\alpha u + \beta \log \cosh u)||_{C^2(\mathbb{R}^1)} \quad \epsilon, \text{ with } \epsilon > 0$$

<u>sufficiently</u> <u>small</u>, <u>then</u> Au <u>is</u> <u>also</u> $C^2$ <u>conjugate</u> <u>to</u> (20.).

### 9. Nonlinear Parabolic Evolution Equations

We consider the equation

$$\frac{du}{dt} + Lu + f(u) = g \ ,$$

$$u\big|_{\partial\Omega} = 0 \ , \tag{21.}$$

subject to Dirichlet data on the boundary of a bounded domain $\Omega$ in $\mathbb{R}^N$, where $L$ is a strongly elliptic linear differential operator. The conventional theory for the equations of this class consists of stydying the Cauchy problem for (21.). This gives rise to the theory of semi-groups and has many successes. However, currently one runs into difficulties with this approach because of chaotic and bifurcation phenomenon. Moreover, the methods of inverse scattering for this problem seem to break down when the space dimension $N > 1$ and when a given nonlinear problem integrable by inverse scattering is slightly perturbed. To overcome this difficulty we shall change the Cauchy problem and consider _periodic_ _solutions_ of the associated parabolic system. This in effect means that we add boundary conditions in the $t$-variable and this will allow us to study bifurcation phenomena from a more conventional point of view.

Thus we add to (21.) the periodic boundary condition reflecting the T-periodicity in the function g on the right hand side of (21.) so that we supplement (21.) by the boundary condition

$$u(0) = u(T) \ . \tag{22.}$$

The next step in our set-up is to translate the problem (21.)-(22.) into a functional analysis setting and to apply the methods of Section 3 and the discussion thereafter. Thus we attempt to define Hilbert spaces $X$ and $Y$ so that the operator $A: X \to Y$ defined by

$$Au = \frac{du}{dt} + Lu + f(u) \ , \tag{23.}$$

subject to the appropriate boundary conditions is a nonlinear operator of Fredholm index zero between the spaces of $X$ and $Y$. In fact we have the following result :

<u>Theorem</u>:   Choose   $X = W_{1,2}(0,T;\Sigma)$ ,

and           $Y = L_2(0,T;\Sigma)$   with   $\Sigma = \overset{\prime}{W}_{1,2}(\Omega)$

then the operator  $A(u)$  defined in equation (23.) is a nonlinear Fredholm operator of index zero acting between the spaces of  X  and  Y.

The next step in our analysis is to analyze the singular points of the mapping  A,  to investigate the bifurcation points and bifurcation values of this mapping and to attempt to classify these bifurcation points as folds and cusps, as defined in the last section.  All this is our future research direction and I think that there are great hopes for success.

Department of Mathematics, University of Massachusetts-Amherst

APPENDIX

Some Proofs: On The Determination of Bifurcation Points

Singular Points of the Riccati operator A

Let $$Ay = y' + y^2, y(0) = y(T).$$

The map $A$ is a Fredholm (nonlinear) operator of index zero between $W_{1,2}$ and $L_2$. Thus the bifurcation points of $A$ coincide with the point $x \in W_{1,2}$ such that $A'(x)y = y' + 2xy$ has a nontrivial kernel. Here we use the fact that $A'(x)y$ has a nontrivial kernel in $X$ with the given boundary conditions if

$$\int_0^T x(t)dt = 0 \qquad (24.)$$

To see the necessity of this condition simply divide the equation

$$y' + 2xy = 0 \qquad (25.)$$

by $y$ and rewrite it as

$$(\log y)' + 2x = 0$$

Then integrate over $(0,T)$. To prove the sufficiency simply write out the explicit solution for equation (25.) and check that (24.) ensures a periodic nontrivial solution.

The same analysis holds for the operator

$$Ay = y' + P_N(y,\lambda).$$

Then

$$A'(x)y = y' + P_N'(x,\lambda)y ,$$

so the necessary and sufficient condition desired is

$$\int_0^T P_N'(x(t),\lambda)dt = 0$$

as stated in the text.

BIBLIOGRAPHY

Articles on the new ideas for complete integrability

1. M. S. Berger and P. T. Church, Complete integrability and perturbation of a nonlinear Dirichlet problem, I. Indiana Univ. Math. J. 28(1979), 935-952. Erratum, ibid 30(1981), 799.

2. M. S. Berger, New ideas on complete integrability. Partial Differential Equations, Banach Center Publications, Volume 10, PWN-Polish Scientific Publishers (1978).

3. M. S. Berger, P. T. Church, and J. G. Timourian, An application of singularity theory to nonlinear elliptic partial differential equations. In "Singularities" (P. Orbits, Ed.), Proc. Symp. in Pure Math. 40 (1983), Part 1, 119-126.

4. M. S. Berger, The diagonalization of nonlinear differential operators-- a new approach to complete integrability. (To appear, Proc. AMS-SIAM Conf. on Nonlinear Partial Differential Equations, 1985).

5. M. S. Berger, P. T. Church, and J. G. Timourian, Integrability of nonlinear differential equations via functional analysis. (To appear, Proc. AMS Conf. on Nonlinear Functional Analysis, 1985).

Articles on infinite dimensional Whitney folds and cusps

1. M. Golobitsky and R. Guillemin, Stable mappings and their singularities. Springer-Verlag, Berlin/Heidelberg/New York (1973).

2. M. S. Berger, P. T. Church, and J. G. Timourian, Folds and cusps in Banach spaces, with applications to nonlinear partial differential equations, I. Indiana Univ. Math. J. Vol. 34, 1(1983), 1-19.

Articles on special equations integrability by new method for the nonlinear elliptic log cosh Gordon equation

1. M. S. Berger, Classical solutions in nonlinear Euclidean field theory and complete integrability, Springer Lecture Notes No. 925 (1982), 123-133.

For the Riccati equation see the article
H. McKean and C. Scovel in Annali di Math. Pisa (to appear, 1986).

FRACTAL SINGULARITIES IN A MEASURE AND HOW TO MEASURE SINGULARITIES
ON A FRACTAL

Reported by

Leo P. Kadanoff
The James Franck Institute
The University of Chicago
5640 S. Ellis Ave.
Chicago, IL 60615

This is a preliminary report upon a piece of research being carried out by Mogens Jensen, Leo Kadanoff, Itamar Procaccia, and Boris Shraiman. It is an outgrowth of the thinking reflected in the work of Hentschel and Procaccia[1] and of Halsey, Meakin, and Procaccia.[2] A fuller report on this work will appear later.

## Singularities

In many different areas of physics one is interested in describing how an object, perhaps fractal, may be covered by a measure. A measure is just a probability assigned to each piece of the object. For example, in dynamical systems theory, the object may be a strange attractor and the measure the probability for visiting a given piece of the attractor. In a random resistor network, the object might be a percolating cluster of resistors and the measure might be proportional to the magnitude of the current through each resistor.

In any case, imagine some object, perhaps fractal, sitting in an ordinary Euclidean space. The object is divided into N pieces with the jth piece having size $l_j$. Associated with the jth piece is $M_j$, the measure of that piece.

This measure may have singularities. For example if the object is the line between -1 and 1 ($x \epsilon [-1,1]$) and the measure between x and x + dx is equal to M(x)dx with $M(x) = |x|^{-1/2}$, then the interval [-1, 1] has measure $\int_{-1}^{1} dx M(x) = 1^{1/2}$. Thus, there is a singularity with index 1/2 at the origin. In general if, for small $l_j$, we find

$$M_j \sim l_j^{\alpha},$$ 

(1)

we say there is a singularity of type $\alpha$.

If the measure were uniformly distributed over a d-dimensional space, $\alpha$ would be d. Hence $\alpha$ is a kind of dimension for the singularity. In ancient times, in critical phenomena, $\alpha$ was denoted by the symbol y.

## Counting Singularities

One example of interest is a DLA aggregate with the measure being the probability that a walker will land at a given point.[3] In this and other examples, there are infinitely many singularities (in the limit as the cluster becomes infinitely large) and a whole range of different possible values of $\alpha$.

To characterize what happens let $N(\alpha)d\alpha$ be the number of singularities of type $\alpha'$ for all $\alpha'$ lying between $\alpha$ and $\alpha + d\alpha$. As we divide the object more and more finely we get more and more singularities. Let l be a typical size of a subdivision and define an index for the number of singularities, $f(\alpha)$, via

$$N(\alpha)d\alpha = \rho(\alpha) \, d\alpha \, l^{-f(\alpha)}. \tag{2}$$

The higher $f(\alpha)$ is, the larger the density of singularities of type $\alpha$. In critical phenomena, $f = d$. Notice that f may be interpreted as the dimension of the set of singularities of type $\alpha$.

## Measuring $f(\alpha)$ and $\alpha$:

Now I wish to know ("measure") the possible values of $\alpha$ and the function $f(\alpha)$ for some particular set. To do this, I follow an approach based upon a partition function

$$Z_N(q,\tau) = \sum_{j=1}^{N} M_j^q \, l_j^{-\tau} . \tag{3}$$

I define $\tau_N(q)$ as the value of $\tau$ which makes this partition function equal to 1:

$$Z_N(q, \tau_N(q)) = 1. \tag{4}$$

I find that the resulting value of $\tau_N(q)$ is not very sensitive to how the splitting into pieces is done and that the limit $N \to \infty$ exists. Hence define

$$\tau(q) = \lim_{N \to \infty} \tau_N(q). \tag{5}$$

Given a data set, $\tau(q)$ can be directly measured

What is the thing that has been measured? Say that, in splitting up the set, we obtain a piece with a given l-value with a probability density $p(l)$. Then from Eqs. (1), (2), and (3) we can write the partition function as

$$Z_N(q, \tau) = \int dl \, p(l) \int d\alpha \, \rho(\alpha) l^{-f(\alpha)} l^{q\alpha} l^{-\tau}. \tag{6}$$

The $\alpha$-integral will contribute almost entirely from the region in which the exponent of l, $q\alpha - f - \tau$, is a minimum. This will occur when

$$q = f'(\alpha), \tag{7a}$$

or else when $\alpha$ reaches an endpoint of a region in which $\rho(\alpha)$ is nonzero. (Here we neglect the latter possibility.) If the result is not to go to zero or infinity as $N \to \infty$, we must have the total exponent of l be zero. (See the condition of Eq. (4)). Thus we also have

$$q\alpha = f(\alpha) + \tau, \tag{7b}$$

as in reference 2.

Our measurement gave us $\tau$ as a function of q. Equations (7) are implicit equations which will express $\alpha$ as a function of q and f as a function of q and will then finally give $f(\alpha)$. To make this conclusion more explicit notice that these equations are of the form of a Legendre transformation. We can rewrite them in the form

$$f(\alpha) = -\tau(q) + q\alpha \, , \tag{8a}$$

where q is determined as a function of $\alpha$ via

$$\tau'(q) = \alpha \, . \tag{8b}$$

## What it Means

Varying q and $\tau$ is a trick for exploring the different regions of $\alpha$. For large positive q we are looking at small values of $\alpha$, i.e., places in which the measure is very highly concentrated. For q large and negative we instead study parts of the fractal for which the measure is very small, i.e., the larger values of $\alpha$. At q = 0, the measure drops out and f is the **Hausdorff** dimension. This turns out to be the maximum possible value of f.

## Acknowledgement

This research reported here was supported by the ONR.

References

1. H. G. E. Hentschel and I. Procaccia, Physica 8D, 440 (1983).

2. T. Halsey, P. Meakin, and I. Procaccia, preprint.

3. This example was studied in detail in reference 2 and also in the parallel and independent work by H. Scher and L. Turkevitch.

4. A. For a mathematical basis for this type of approach see R. Bowen, Equilibrium States and the Ergodic Theory of Anisov Diffeomorphisms, Lecture Notes in math. No. 470, Springer, Berlin, 1975. B. For a somewhat analogous attack upon random resistor networks see L. de Arcangelis, S. Redner, and A. Coniglio, Phys. Rev. B31, 4725 (1985).

# LÉVY WALK REPRESENTATIONS OF DYNAMICAL PROCESSES

by

Michael F. Shlesinger
Office of Naval Research
Physics Division
800 North Quincy Street
Arlington, Virginia   22217

and

Joseph Klafter
Corporate Research Science Laboratory
Exxon Research & Engineering Company
Annandale, New Jersey   08801

## ABSTRACT

A stochastic process, called a Lévy walk, is introduced.  For different scaling assumptions this process provides a statistical description of accelerated diffusion in Josephson junctions and diffusion in a turbulent fluid.  The accelerated diffusion case has also been successfully described by a one dimensional deterministic nonlinear mapping while the case of turbulent diffusion has so far not been generated from a deterministic map.

## I.  INTRODUCTION

The discovery of <u>dynamical</u> chaos has led some to predict the demise, or at least the downgrading, of <u>statistical</u> mechanics and <u>stochastic</u> processes as viable or useful descriptions of nature.  We, on the other hand, feel that the complexities arising from dynamical equations of motion present new challenges to statistical physics.  Some obvious examples are calculating diffusion constants for phase space motion in Hamiltonian systems which possess intrinsic stochasticity, predicting the asymptotic behavior of dissipative systems with fractal basin boundaries, and providing a proper statistical description of motion governed by strange attractors especially in high dimensions.  From our perspective the interplay and mixing between nonlinear dynamics and stochastic processes will be very fruitful and exciting area of theoretical study.

In this manuscript we will discuss statistical representations of two nonlinear processes, accelerated phase diffusion in Josephson junctions and turbulent diffusion in fluids.  Section II reviews the work of Geisel, Nierwetberg, and Zacherl[1] on accelerated diffusion in chaotic systems. Section III presents a stochastic process called a Lévy walk[2,3] with the same statistical properties of accelerated diffusion found in Section II. A change of scaling in the Lévy walk is shown in Section IV to reproduce the even more rapid diffusion described by Richardson's equation for turbulent diffusion[4-6].

## II.  ACCELERATED DIFFUSION IN CHAOTIC SYSTEMS

It was generally thought that diffusion, described by

$$\langle x^2(t) \rangle = Dt \qquad (1)$$

was a property only of random systems.  It is now known that such behavior can result for motion induced by deterministic mappings.  For example, the map[7]

$$x_{t+1} = x_t - \mu \sin(2\pi x_t) \qquad (2)$$

produces diffusive motion, $\langle(x_t-x_0)^2\rangle \sim Dt$ (the angular brackets indicate an average over initial conditions) when $\mu > \mu_{critical} \sim 0.732$. Mappings[8] can also produce anomalous diffusion where the diffusion constant is rigorously zero, i.e. $\langle(x_t-x_0)^2\rangle \sim t^\alpha$, $\alpha < 1$. This occurs if the map $f(x_t)$ closely approaches the 45° line in the plot of $x_{t+1}$ vs. $x_t$, so the trajectory gets stuck for many iterations in the same region for a long time $T(x_0)$. This idea was introduced by Pomeau and Manneville[9] in their dynamical theory of intermittency, and used by Geisel and Thomae[8] to find the universal z behavior

$$\langle(x_t-x_0)^2\rangle \sim \begin{cases} t^{\frac{1}{(z-1)}} & , z > 2, \\ \\ t & , 1 < z < 2, \end{cases} \tag{3}$$

for maps with the following behavior near the origin,

$$x_{t+1} = x_t + ax_t^z \quad \text{for } x_t \to 0^+, \tag{4}$$

with $a > 0$ and $z > 1$. The slower than diffusive behavior results because some initial conditions lead to trajectories which spend extremely long times in the region near the origin. Geisel, Nierwetberg, and Zacherl[1] have extended this analysis to mappings which have a near tangent to the 45° line, but in a neighboring unit cell at its boundary, (see Fig. 1). When the trajectory leaves the unit cell this represents a motion to a neighboring cell. Periodic boundary conditions bring the new initial position back to the original unit cell. In the previous analysis the time $T(x_0)$ spent near the origin now becomes the time spent in correlated (constant velocity) motion moving between unit cells. The mapping[1]

$$x_{t+1} = f(x_t),$$

where

$$f(x) = (1+\epsilon)(x-m) + a(x-m)^z + m-1 \quad (m \leq x < m+1/2). \tag{5}$$

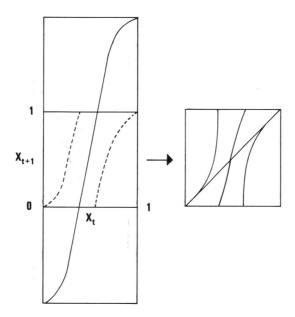

FIG. 1.    The mapping of Geisel, Nierwetberg, and Zacherl.  The trajectory
gets trapped for many iterations in the upper right and lower
left hand corners and is transferred out of the unit cell
successively resulting in a correlated motion that leads to
accelerated diffusion for z > 3/2.

for $(x-m) \ll 1$ and $\varepsilon \to 0$ when averaged over all initial positions leads to

$$\langle (x_t - x_0)^2 \rangle \sim \begin{cases} t^2, \ z > 2, \\[2mm] t^{3 - \frac{1}{(z-1)}}, \ \frac{3}{2} < z < 2, \\[2mm] t \ \ln t, \ \ z = 3/2, \\[2mm] t, \ \ \ 1 < z < 3/2. \end{cases} \qquad (6)$$

The cases for z > 3/2 are called accelerated diffusion.  The mapping Eqn.
(5) has been used to approximate the complex voltage phase behavior in
resistively shunted Josephson junctions.  In the next section we present a
stochastic process with precisely the same behavior of Eqn. (6).

## III. LÉVY FLIGHTS AND WALKS[2,3,10]

Our main concern here is to present a stochastic process which has the same four types of behavior found for the dynamical system in Eqn. (6). We first analyze a <u>Lévy flight</u>. We use this term to denote a random walker which jumps instantaneously between successively visited sites, however distant. The preface Lévy to the flight implies that the distribution of jump distances has an infinite second moment. The term <u>Lévy walk</u> describes the same random walk but with the walker forced to visit the intervening sites between successively visited sites of the Lévy flight. In this case the jump is not instantaneous, but has a finite velocity. See Figure 2. However, different length jumps may occur at different velocities.

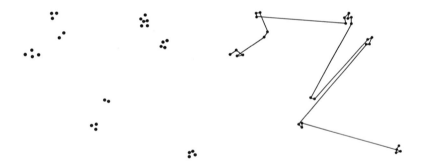

FIG. 2.    The set of points on the left are those visited by a random Lévy flight and are characterized by a fractal dimension β. The points on the right are the same as those on the left, except that they are connected by straight lines. The exponent γ determines the velocity with which the straight lines are traversed.

To simplify the analysis of these stochastic processes, we limit our discussion to a one dimensional periodic lattice.

Let $p(x)$ be the probability that a random walker, in a single step, has a displacement $x$. Also denote by $P_n(x)$ the probability that the walker, if originally at the origin, is at site $x$ after $n$ steps. Then

$$P_{n+1}(x) = \sum_{x'} P_n(x-x') \, p(x') . \tag{7}$$

Fourier transforming over the lattice sites, we find using the convolution theorem, that

$$\tilde{P}_n(k) = \sum_x e^{ikx} P_n(x) = [p(k)]^n ,$$

or equivalently,

$$P_n(x) = \frac{1}{2\pi} \int_0^{2\pi} [p(k)]^n e^{-ikx} dk. \tag{8}$$

For jumps restricted to nearest neighbors with equal probability,

$$p(x) = \frac{1}{2} [\delta_{x,+1} + \delta_{x,-1}]$$

and

$$p(k) = \cos k \sim 1 - \frac{1}{2} k^2 \text{ as } k \to 0 . \tag{9}$$

For any unbiased jump distribution, with a finite variance $\overline{x^2} \equiv \sum_x x^2 p(x)$, one can always expand $p(k)$ as in Eqn. (9) :

$$p(k) = \sum_x e^{ikx} p(x) \sim 1 - \frac{1}{2} \overline{x^2} k^2 \text{ as } k \to 0$$

$$\sim \exp(-\frac{1}{2} \overline{x^2} k^2) . \tag{10}$$

Substituting this $p(k)$ in Eqn. (8) for $P_n(x)$ yields the standard gaussian behavior,

$$P_n(x) \sim (2\pi n \, \overline{x^2})^{-1/2} \exp(-x^2/2n \, \overline{x^2}) , \tag{11}$$

in accord with the Central Limit Theorem.

When the vogue in probability theory was to see what were the weakest conditions under which gaussian behavior would result, P. Lévy sought exceptions to the Central Limit Theorem. He eventually considered random walk processes where $\overline{x^2}$ was infinite, so the trajectory possessed no largest length scale. If for large x,

$$p(x) \sim |x|^{-1-f}, \text{ with } 0 < f < 2$$

then $\overline{x^2}$ is infinite, and[10]

$$p(k) \sim 1 - \text{const.} \ |k|^f \quad \text{as } k \to 0$$
$$\sim \exp(-\text{const.} \ |k|^f) \tag{12}$$

and

$$P_n(x) \sim \frac{1}{2\pi} \int_0^{2\pi} \exp(-\text{const.} \ |k|^f) \ \exp(-ikx)dk. \tag{13}$$

The probability $P_n(x)$ is known as a Levy density and its integral representation can only be calculated analytically for a few select cases, e.g., f = 1 corresponds to

$$P_n(x) = \frac{1}{\pi} \frac{An}{x^2 + (An)^2} , \tag{14}$$

where A is a constant. The exponent f can be shown[10] to be the fractal dimension of the set of points visited by the random walker. For the Lévy flight governed by Eqns. (12) and (13), the mean square displacement

$$\langle x^2(t) \rangle \equiv \sum_x x^2 P(x,t) = \infty , \tag{15}$$

since $\overline{x^2}$ is infinite for a single jump. How can Eqn. (15) be transformed to reproduce the mapping results of Eqn. (6)?

A manner in which to obtain a finite time dependent result for $\langle x^2(t) \rangle$ is to account for the time taken to move between the initial and final positions of a jump. We will assume that the random walker moves with a constant velocity, i.e. longer jumps take a longer time. Different size jumps may have different velocities. To properly account for this feature we must generalize Eqn. (7) to

$$Q(x,t) = \sum_{x'} \int_0^t Q(x-x', t-\tau) \, \Psi(x',\tau) \, d\tau + \delta_{x,0} \delta(t) \, , \qquad (16)$$

where $Q(x,t)$ is the probability for reaching site x <u>exactly</u> at time t for a random walker starting at the origin at $t = 0$. The quantity $\Psi(x,t)$ is the probability density for the random walker to reach site x at time t in a single jump. Thus one can reach site x at time t by first reaching site $x-x'$ at time $t-\tau$ (accounting for all possible pathways) and then in the remaining time $\tau$ the random walker makes a jump of length $x'$. All possible intermediate sites are summed over.

We write the memory function $\Psi(x,t)$ as

$$\Psi(x,t) = \psi(t|x)p(x) \, , \qquad (17)$$

where $p(x)$ is the probability of making a jump of length x, or in a different interpretation making x correlated jumps of unit length all in the same direction. We further write

$$\psi(t) = \sum_x \Psi(x,t) \, , \qquad (18)$$

where $\psi(t)dt$ is the probability that if a jump ended at time $t = 0$ then the next jump terminates in the time interval $(t, t+dt)$. The probability $P(x,t)$ for being at site x at time t is somewhat different from $Q(x,t)$ because the walker might reach site x at an earlier time $t-\tau$ and then not move for a time $\tau$, i.e.

$$P(x,t) = \int_0^t Q(x, t-\tau) \int_\tau^\infty \psi(t') \, dt' \, d\tau . \qquad (19)$$

Both Fourier and Laplace transforming Eqn. (19) we find

$$\sum_x \int e^{-st} e^{ikx} P(x,t)dt = \tilde{P}*(k,s) = \tilde{Q}*(k,s) \frac{1-\psi*(s)}{s} \, , \qquad (20)$$

and from Eqn. (16)

$$\tilde{Q}*(k,s) = \frac{1}{1-\tilde{\Psi}*(k,s)} . \qquad (21)$$

For a random walk without a bias,

$$\int_0^\infty e^{-st} \langle x^2(t) \rangle dt \equiv \sum_x x^2 p*(x,s)$$

$$= -\frac{\partial^2}{\partial k^2} \tilde{p}*(k = 0, s)$$

$$= \frac{1}{s[1-\psi*(s)]} \frac{\partial^2}{\partial k^2} \tilde{\Psi}*(k = 0, s) . \tag{22}$$

Let us choose

$$p(x) \sim \frac{1}{|x|^{2+\beta}} \text{ as } x \to \infty \tag{23a}$$

and
$$\psi(x|t) = \delta(|x|-t) . \tag{23b}$$

Note $\psi*(s) = \sum_x \Psi*(x,s) = \int_0^\infty \exp(-st)p(t)dt.$

**For $0 < \beta < 1$,**

$$\psi*(s) \sim 1-s\langle t \rangle + \text{const. } s^{1+\beta} \text{ as } s \to 0 , \tag{24}$$

where $\langle t \rangle \equiv \int_0^\infty t\psi(t)dt$ is the average time spent on a jump. For $-1 < \beta < 0$, $\langle t \rangle$ is infinite and

$$\psi*(s) \sim 1 - \text{const. } s^{1+\beta}, \text{ as } s \to 0. \tag{25}$$

For these two $\beta$ regimes, $-1 < \beta < 1$,

$$\partial^2 \tilde{\Psi}*(k = 0, s)/\partial k^2 \sim s^{\beta - 1} \tag{26}$$

Substituting Eqns. (23) - (26) in the expression for the mean square displacement, Eqn. (22), gives

$$\langle x^2(t) \rangle \sim \begin{cases} t^2, \ -1 < \beta < 0, \ \langle t \rangle = \infty, \\[2mm] t^{2-\beta}, \ 0 < \beta < 1, \ \langle t \rangle < \infty, \ \langle t^2 \rangle = \infty, \\[2mm] t \ \text{lnt}, \ \beta = 1, \ \langle t^2 \rangle \ \text{log divergent}, \\[2mm] t, \ \beta > 1, \ \langle t^2 \rangle < \infty. \end{cases} \tag{27}$$

This is the equivalent of the dynamical result if we set

$$\beta = \frac{2 - z}{z - 1}. \tag{28}$$

The $t^2$ behavior originates from the first moment of $\psi(t)$ diverging, the next two cases originate from $\langle t \rangle$ finite, but $\langle t^2 \rangle$ infinite, and the standard Brownian motion result corresponds to $\langle t^2 \rangle$ finite.

In the Josephson junction mapping if the voltage phase rotates, say, counterclockwise, n times, this corresponds to the Lévy walker taking n correlated steps, say to the right, while clockwise rotations would correspond to correlated steps to the left. The $t^2$ result should not be misinterpreted to mean that wave motion is occurring. In fact, the motion is quite random and involves correlated jumps over an infinite number of scales. The meaning of $\langle t \rangle$ being infinite is that there is no largest characteristic time scale (and in this case also correlated jump distances have no largest scale), as can be seen in the following pedogogical example.

Choose

$$\psi(t) = \frac{1-N}{N} \sum n^j b^j \exp(-tb^j), \ N, b < 1, \tag{29}$$

where jumps occur on an infinite of time scales $b^{-1}$, $b^{-2}$, $b^{-3}$, . . . .etc., but an order of magnitude longer time scale in base $b^{-1}$ occurs with an order of magnitude less probability in base $N^{-1}$. About $N^{-1}$ events occur on a time scale of $b^{-1}$ before an intermittent interval of length $b^{-2}$ occurs. Then about $N^{-1}$ events separated in time by an interval $b^{-1}$ occur again and this happens about $N^{-1}$ times before an intermittent interval of length $b^{-3}$

occurs, and so on with intermittences of all time scales entering in a self-similar manner. These intermittent intervals can represent the time a random walker spends in a correlated motion, and one expects the set of event times to have a fractal dimension of log N/log b. To see this analytically, let us Laplace transform Eqn. (29) to obtain

$$\psi*(s) = \frac{1-N}{N} \sum_{j-1}^{\infty} \frac{(Nb)^j}{b^j + s} = N\psi*(s/b) + (1+N) \frac{b}{b+s} \cdot \tag{30}$$

The homogeneous part of this equation,

$$\psi*(s) = N\psi*(s/b) ,$$

has the solution

$$\psi*(s) = s^f K(s),$$

where

$$f = \ln N/\ln b$$

and K(s) is a function periodic in ln s with period ln b. It can be shown[10] that f is the fractal dimension of the set of times marking the switching between different rotation states.

IV.  RICHARDSON'S LAW

At first glance it appears that $\langle x^2(t) \rangle = t^2$ is the fastest possible motion for the Levy walker because it involves an infinite correlation length. This is true if all the jumps occur at the same constant velocity. If however, a distribution of velocities exists then one can arrive at a more rapid diffusion which is associated with turbulence.

Let us choose in contrast to Eqn. (23b),

$$\psi(x|t) = \delta(|x|-t^\gamma) , \tag{31a}$$

$$\text{with } \gamma = 3/2, \tag{31b}$$

but keep the same form of p(x) as in Eqn. (23a).

Analysis similar to the calculations of Section III show that if

$$-1 < \beta < -1/3 ,$$

then

$$\langle x^2(t) \rangle \sim t^3 . \tag{32}$$

Eqn. (32) was given by Richardson in 1926 to describe turbulent diffusion. In our approach we envision a random walker in a distribution of vortices. The distribution of correlation lengths $p(x)$ corresponds to the distribution of vortex sizes. Larger vortices have more energy and a larger velocity so the time t to complete a correlated motion of length x is proportional to $x^{2/3}$ rather than x as in the Josephson junction accelerated diffusion problem.

Richardson suggested that

$$\frac{\partial P(x,t)}{\partial t} = \frac{\partial}{\partial x} K(x) \frac{\partial P(x,t)}{\partial x} , \tag{33}$$

with $K(x) = $ const. $x^{4/3}$. This scaling leads to

$$\frac{d\langle x^2(t) \rangle}{dt} = [\langle x^2(t) \rangle^{1/2}]^{4/3} , \tag{34}$$

Richardson's famous "4/3" law which is equivalent to Eqn. (32). Our random walk picture is equivalent to a non-local in space and time master equation,

$$\frac{dP(x,t)}{dt} = \int \sum_{x'} (P(x-x') \ t-\tau) \ \Phi(x',\tau)$$

$$- P(x,t-\tau) \ \Phi \ (x',\tau))d\tau , \tag{35}$$

where[10]

$$\frac{\Phi*(k,s)}{s + \tilde{\Phi}*(k=0,s)} = \tilde{\Psi}*(k,s) . \tag{36}$$

Correction to Richardson's law due to the breakdown of the concept of
homogeneous turbulence because the turbulence is concentrated on a fractal
set $d_f$ has been suggested by Mandelbrot[11]. If D is the euclidean dimension
then it has been concluded[12] that the Richardson 4/3 is replaced by (4+2(D-
$d_f$))/3. Such corrections can be incorporated into the Lévy walk by
choosing $\gamma > 3/2$ in Eqn. (31b). Then if $\gamma (1+\beta) < 2$ the behavior

$$\langle x^2(t)\rangle \sim t^{2\gamma} \tag{37}$$

ensues asymptotically. Several slower behaviors, down to the Brownian
motion result, can be obtained for different choices of $\gamma$ and $\beta$, as we saw
in the equation for the accelerated diffusion example in Eqn. (27). Much
further work remains to be completed for the statistical analysis of a **Lévy**
walk restricted to various fractal sets.


CONCLUSION

The Lévy walk with a constant velocity was able to reproduce the
statistical behavior of accelerated diffusion which has also been derived
from a deterministic one dimensional map. Lévy walks with a distribution
of velocities correlated to vortex sizes can generate the proper mean
square displacement of a particle in a turbulent fluid. It would be of
interest to explore if a dynamical map could also achieve this last result.

REFERENCES

1. T. Geisel, J. Nierwetberg, and A. Zacherl, Phys. Rev. Lett. 54, 616
(1985).
2. M. F. Shlesinger and J. Klafter, Phys. Rev. Lett. 54, 2551 (1985).
3. M. F. Shlesinger and J. Klafter, in Proceedings of the Cargèse Summer
School, "On Growth and Form: A Modern View", eds. H. E. Stanley and N.
Ostrowski (in press).
4. L. F. Richardson, Proc. Roy. Soc. London A110, 709 (1926).
5. H. G. E. Hentschel and I. Procaccia, Phys. Rev. A27, 1266 (1983).
6. F. Wegner and S. Grossman, Zeits. für Physik B59, 197 (1985).

7.  T. Geisel and J. Nierwetberg, Phys. Rev. Lett. 48, 7 (1982).

8.  T. Geisel and S. Thomae, Phys. Rev. Lett. 52, 1936 (1984).

9.  Y. Pomeau and P. Manneville, Comm. Math. Phys. 74, 189 (1980).

10. E. W. Montroll and M. F. Shlesinger, in Studies in Statistical Mechanics, Vol. 11, eds. J. Lebowitz and E. W. Montroll, (North-Holland, Amsterdam) 1984.

11. B. B. Mandelbrot, on Turbulence and the Navier-Stokes Equation, ed. R. Temam (Springer, Berlin, 1976).

12. H. G. E. Hentschel and I. Procaccia, Phys. Rev. Lett. 49, 1158 (1982).

B. A. Nelson and J. Hermanson, Phys. Rev. Lett. _40_, 1115 (1978).

E. J. Baerends and E. Snijders, J. Chem. Phys. _71_, 3289 (1979).

J. V. Tucker and R. Erdahl, Phys. Lett. _46A_, 149 (1978).

C. F. Moore and R. C. Morrison, Int. J. Quantum Chem. _5_, 177 (1971).

U. Manthe and T. Weil, J. Phys. B: At. Mol. Opt. Phys. _12_, 145 (1979).

R. P. Messmer, Phys. Rev. _B4_, 1547 (1971).

J. A. Hermanson, in Proc. Int. Conf. on Physical State Physics, ...

K. Terem, Physics Reports _39C_, 135 (1978).

David D. J. Keith and J. Freed, Z. Phys. Rev. _B25_, 1851 (1982).